SpringerBriefs in Materials

For further volumes:
http://www.springer.com/series/10111

Pritam Deb

Kinetics of Heterogeneous Solid State Processes

 Springer

Pritam Deb
Department of Physics
Tezpur University (Central University)
Tezpur, Assam
India

ISSN 2192-1091   ISSN 2192-1105 (electronic)
ISBN 978-81-322-1755-8   ISBN 978-81-322-1756-5 (eBook)
DOI 10.1007/978-81-322-1756-5
Springer New Delhi Heidelberg New York Dordrecht London

Library of Congress Control Number: 2013954033

Printed on acid-free paper

Springer is part of Springer Science+Business Media (www.springer.com)

Preface

Research and development of kinetic studies has become one of the most expanding fields in physics, chemistry and materials science. One reason for this trend is that this field bridges various scientific disciplines. The hype concerning nanotechnology in recent years has given an additional boost to this topic. The requirements for better understanding of this fundamental aspect is a driving force for the research and development in this area.

This book provides scientists, researchers and also interested people from other branches of science, with the opportunity to learn a new method of non-isothermal kinetic analysis and its application in heterogeneous solid state processes.

The book is divided into six chapters: Chap. 1 is an introduction to the basic concepts of kinetics; Chap. 2 describes a new, realistic and more accurate non-isothermal kinetic method; Chap. 3 shows application of this method on the mechanistic determination of evolution of a nanosystem and Chap. 4 is kinetic analysis of a heterogeneous solid state process through the aforementioned method.

The breadth of the topic means that not all topics can be covered; however the interested reader will find additional references at the end.

It is a great pleasure to thank those who have helped during the course of writing this book. I wish to record my gratitude to my students who have sometimes made me think very hard about things I thought I understood. I have benefitted greatly from discussions with Prof. Amitava Basumallick and his suggestions. I am also grateful to my student Kakoli Bhattacharya for fruitful discussions and technical assistance. I am leaving the book in your hands and take your leave sharing a quotation from one of the greatest philosophers, Oliver Wandell Holmes, who said "Man's mind, once stretched by a new idea, never regains its original dimensions". It is my hope that this book will help in stretching the limits of thinking in all those who come into contact with it.

Tezpur, December 2013 Pritam Deb

Contents

Contents

Chapter 1
Fundamental Concepts of Kinetics

The solid state reaction kinetics has been extensively studied time and again in the past century. The parameters involving in the kinetics of solid state processes are often misconstrued as intrinsic constants that characterize a particular solid state reaction. But this is a delusion as solid state processes involve rigorous and complex kinetics. The study of the kinetics in the solid state explicates the mechanism of the process as well as the related kinetic parameters. Kinetic studies have traditionally been extremely useful in characterizing several physical and chemical phenomena in organic, inorganic and metallic systems, which involve thermal effects. It provides valuable qualitative, quantitative information on the thermal properties and, more importantly, kinetic information on phase transformations [1], crystallization of metallic glasses [2], solid state precipitation [3, 4], decomposition [5], quasicrystalline structure of amorphous metals and alloys [6] and structural changes associated with nanomaterials [7]. However, it is to be noted that the physical properties of the materials/particles do not solely depend on the composition, but are also functions of shape and size. It has been reported that deviation in shape and size of the nanoparticles impart a great deal of inconsistency in the results of physical properties [8]. Therefore, it is strongly felt that unless the physical properties are tailored to remain uniform through precise control of shape and size, the acceptance and applicability of the materials/ particles appears to be eclipsed. Fundamentally, realization of such control requires in-depth knowledge about the reaction mechanism and the related kinetic parameters, e.g. temperature, heating rate, etc.

Kinetics and thermodynamics are two important aspects for understanding the solid state processes. Thermodynamics is concerned about the initial and final states of a process, whereas kinetics additionally deals with mechanism, reaction path and the time required. Thermodynamics always deals with equilibrium condition. Hence, kinetic study in addition to thermodynamics is required for characterizing solid state process.

All the particles in a system do not undergo transformation at one and the same time. Only a fraction of the active particles can participate in the transformation process because they only have the free energy of the excess of the mean and hence are energetically suitable for transformation. The free energy of an atom or

P. Deb, *Kinetics of Heterogeneous Solid State Processes*, SpringerBriefs in Materials, DOI: 10.1007/978-81-322-1756-5_1, © The Author(s) 2014

Fig. 1.1 Energy of the
reacting atoms in an
exothermic reaction

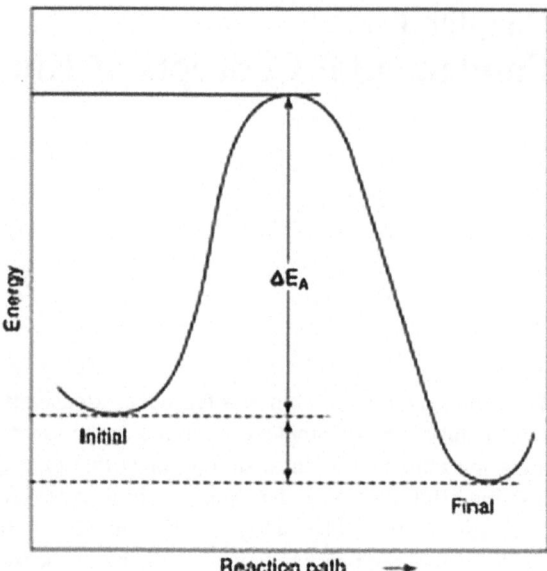

group of atoms during transformations first increases to a maximum and then
decreases to a final value. The process is called activation and the state associated
with the maximum energy is called the activation state. The activation energy is
defined as the difference between the internal energy of the system in the transition
state and that in the initial state (Fig. 1.1).

To evaluate the kinetic behaviour of solid state processes, generally, two distinct methods are employed: first, the yield-time measurements are made while
keeping the reactant species at a constant temperature; secondly, the reactant
molecules are treated under a controlled increasing temperature. The isothermal
method is the only method used earlier for kinetic studies and hence the concepts
of solid state kinetics were developed on the basis of experiments carried out under
isothermal conditions. In the isothermal method, the reactant species are maintained at a constant temperature throughout the reaction period which is not truly
feasible as a finite time required to heat the material to reaction temperature.
Isothermal methods are similar to those used in homogeneous kinetics to produce
α-time data compared to concentration–time data in homogeneous kinetics.

1.1 Activation Energy

For any chemical reaction to take place there is a need for collision between the
reactants. This collision caters to the energy required for the reactants to undergo a
transformation and finally yielding in the desired product. The term activation
energy, coined by the Swedish scientist Svante Arrhenius in 1889, is assigned to

that minimum amount of energy required to initiate a chemical reaction. In case of solid state processes, reactions undergo a spontaneous rearrangement of atoms into new and more stable atomic arrangements. For a reaction to go from the unreacted to the reacted state, in a solid state process, the reacting atoms must possess a certain value of energy to overcome an energy barrier. This additional energy required by the atoms is the activation energy ΔE_a and is usually given in units of joules per mole or calories per mole.

Atoms in the initial state of the reaction possess an average energy E_r (energy of the reactants), which when supplied with the activation energy ΔE_a, is sufficient to cross the energy barrier and transform into the desired product of energy E_p. The activation energies of the atoms is a function of temperature, as at a given temperature only a certain fraction of atoms or molecules will have enough energy to reach the activation energy level. With the increase in temperature, the number of atoms reaching the activation energy level will increase drastically. The effect of temperature on the increase in the energies of the molecules has been described by Boltzmann, where the probability of finding an atom or molecule in the activation energy state E_a greater than the average reactant energy of all the atoms or molecules of a species at temperature T is given by $P \propto e_a^{-(E - E)/kT}$, where k is Boltzmann's constant.

1.2 Arrhenius Law

Arrhenius law gives the dependence of the rate constant of the reactions on the temperature and the activation energy. It was proposed by Svante Arrhenius in 1889, soon after Van't Hoff formulated the van't Hoff equation which relates the equilibrium constant, K, of a reaction, to the heat of the reaction ΔH_T.

$$\frac{dlnK}{dT} = \frac{\Delta H_T}{RT^2} \tag{1.1}$$

As the equilibrium constant K is the ratio of the rate constants k1 and k − 1 in the reverse directions, these constants follow the equation

$$\frac{dlnk}{dT} = \frac{E}{RT^2} \tag{1.2}$$

where, the value of E might depend upon temperature. Thus,

$$k = A \, exp \, (-E/RT) \tag{1.3}$$

which is known as the Arrhenius equation. Here, A is the pre-exponential factor or simply the prefactor, k the rate constant, T the temperature in Kelvin, R ($R = Nk_B$ where N is Avogadro number) the gas constant and E the activation energy.

The van't Hoff equation reveals the exponential dependence of the reaction rate on temperature. Arrhenius, while working on the hydrolysis rate of sugar cane by mineral acids [9], found that the temperature dependence of the rate cannot be determined by the translational energy of molecules or the viscosity of the medium. He discovered that in a reaction, there are both active as well as inactive molecules which are always in equilibrium and this equilibrium is exponentially dependent on the temperature, as predicted by Van't Hoff in his study.

A plot of ln$k$ against reciprocal temperature should thus yield a straight line with its slope as activation energy. This expression has been found to fit experimental data well over wide temperature ranges and validated as a reasonable first approximation in most processes. Not all reactions show, for the rate constant, the Arrhenius type variation. The exceptional cases, often known as the anti-Arrhenius type reactions generally involve changes in the reaction mechanism with temperature.

1.3 Model Fitting Approach

Rate law of a solid state reaction was expressed by a simple differential equation given by:

$$\frac{d\alpha}{dt} = k(T). f(\alpha) \qquad (1.4)$$

where, k(T) is the dependent rate constant, α is the fractional conversion and f(α) is the reaction model. This reaction model may take up various forms corresponding to the various mechanisms of solid state processes.

Different models have been derived from the assumption of the simplified geometry of the diffusion process and its nucleation and growth process. Solid state reaction is generally an amalgamation of various physical and chemical processes which include solid state decomposition, reaction of gaseous product with solid, sublimation, diffusion, adsorption, desorption, melting, etc. Hence, the activation energy of a reaction is basically a composite value determined by the activation energies of various processes as well as the relative contribution of these processes to the total reaction rate. Therefore, activation energy is generally a function of temperature. In addition, in isothermal conditions, the relative contribution of the elementary steps to the overall reaction rate varies with the extent of conversion, resulting in the dependence of the effective activation energy on the extent of reaction [10].

Non-isothermal method involves the heating of the sample at a constant rate. The non-isothermal reactions are more convenient to carry out than the isothermal reactions as in the former case there is no necessity for a sudden temperature jump of the sample at the beginning. It has been seen that the Arrhenius parameters obtained from a non-isothermal study are often reported to disagree with the values obtained from isothermal experiments. There have been two general reasons for this discrepancy, as stated by Vyazovkin and Wight [11]. The first reason is due to the prevalent use of kinetic methods that involve force fitting of non-isothermal data to hypothetical reaction models. Following this "model-fitting approach", Arrhenius parameters are determined by the form of f(α) assumed. Because in a non-isothermal experiment both T and α vary simultaneously, the model-fitting approach generally fails to achieve a clean separation between the temperature dependence, k(T), and the reaction model, f(α). As a result, almost any f(α) can satisfactorily fit data at the cost of drastic variations in the Arrhenius parameters, which compensate for the difference between the assumed form of f(α) and the true but unknown reaction model. For this reason, the model-fitting methods tend to produce highly uncertain values of Arrhenius parameters.

The second major reason for this disagreement arises from the fact that isothermal and non-isothermal experiments are necessarily conducted in different temperature regions. If decomposition involves several steps with different activation energies, the contributions of these steps to the overall decomposition rate measured in a thermal analysis experiment will vary with both temperature and extent of conversion. This means that the effective activation energy determined from thermal analysis experiments will also be a function of these two variables. However, the usual implementation of model-fitting methods is aimed at extracting a single value of the activation energy for an overall process. The value obtained in such a way is in fact an average that does not reflect changes in the reaction mechanism and kinetics with the temperature and the extent of conversion. The aforementioned drawbacks of model-fitting can be avoided with the use of isoconversional methods. First, these methods allow the activation energy to be determined as a function of the extent of conversion and/or temperature. Secondly, this dependence is determined without making any assumptions about the reaction model. Because the model-free isoconversional methods eliminate the causes of the aforementioned disagreement, they are likely to produce consistent kinetic results from isothermal and non-isothermal experiments.

1.4 Isoconversional Method

Model fitting approaches have been criticized in the solid state kinetics of non-isothermal studies because they depend on a constant kinetic triplet: A, E_a and model. This approach involves the fitting of these three parameters which are simultaneously determined from a single curve. They involve a single heating rate which is not always sufficient to determine reaction kinetics. Because of all these drawbacks, isoconversional methods have gained importance. In the isoconversional system, the activation energy is calculated at progressive degrees of conversion (α) without any modelled assumptions [11, 12].

Arrhenius equation relates the rate constant of one simple state reaction through a simple one-step reaction to temperature through the activation energy (E_a) and pre-exponential factor (A). It has been traditionally assumed that E_a and A remain constant, however, it has been shown that in some solid-state reactions, these kinetic parameters may vary with the progress of the reaction (α). This variation can be detected by isoconversional methods.

Isoconversional methods rely on several TGA or DSC datasets for kinetic analysis. When performing non-isothermal experiments, care must be taken to ensure that each run is conducted under the same experimental conditions (i.e., sample weight, purge rate, sample size, particle morphology, etc.) so that only the heating rate varies for each run. For example, sample mass varying from one run to another may cause:

a. Variation in endothermic or exothermic effects (i.e., self-heating or self-cooling), inducing deviations from a linear heating rate.
b. Variation in diffusional rates of evolved gases.
c. Thermal gradients varying with sample mass.

Similarly, sample packing could affect solid-state reaction kinetics where loosely packed powders have large gaps that may reduce thermal conductivity or trap evolved gases compared to a more densely packed powder, which would minimize these effects. Uncontrolled experimental conditions could cause a thermogram to be altered such that it falls above or below its expected location for a non-isothermal study. This results in errors in the calculation of the activation energy by isoconversional methods, which are manifested by a false or artifactual variation in activation energy.

1.5 Kissinger Method

Kissinger relationship, the most extensively used method in kinetic studies since 1957 [13], was in use to determine the energy of activation and the order of reaction, from plots of the logarithms of the heating rate against the temperature inverse at the maximum reaction rate in isothermal conditions. This method is usually based on the Differential Scanning Calorimetry (DSC) analysis of formation or decomposition processes and in relation to these processes the endothermic and exothermic peak positions are related to the heating rate.

According to Kissinger,

$$\ln \beta/T_m^2 = \ln\left[AR\left\{n(1 - \alpha)_m^{n-1}\right\}/E_a\right] - E/RT_m \qquad (1.5)$$

The activation energy, E, is obtained by plotting the left-hand side of the above equation versus $1/T_m$ for a series of runs at different heating rates. However, the method does not calculate E values at progressive α values, but rather assumes a constant E, like other methods.

1.6 Physical Parameters Affecting the Solid State Kinetics

Various steps comprising the solid state kinetics as predetermined by Paul and Curtin [14] are:

- Loosening of molecules at the reaction site
- Molecular change
- Solid solution formation
- Separation of the product phase.

This four-step hypothesis plays a vital role in providing a better insight into the solid state reactions. As in case of reactions in the solution phase, the solid state reactions are also affected by temperature. The temperature dependence on the reaction kinetics can be obviated from the Arrhenius equation. The solid state reactions usually demands a higher activation energy in comparison to the reactions in solution, as in solid state the molecules are bound with a greater energy and as a result there is a higher constraint in the crystal lattice. Time and again, the applicability of the Arrhenius equation has been contested by many but it has been generally observed in narrow temperature ranges. Garn has stressed that the Arrhenius equation is meaningfully applicable only to reactions that take place in a homogeneous environment. However, the Arrhenius equation has been quite successful in describing the temperature dependence of many thermally activated physical processes such as nucleation and growth or diffusion, presumably because the system must overcome a potential energy barrier, and the energy distribution along the relevant coordinate is governed by Boltzmann statistics. Nevertheless, a change in temperature beyond that range would default the validity of Arrhenius equation. Moreover, reaction mechanisms will change with the variation in temperature. Therefore utmost caution should be taken in order to determine the effect of temperature on the kinetics of solid state reactions.

The first step of such solid state reactions involves the loosening of molecules at reaction site which implies that the site of initiation of solid state reactions may have some amorphous characteristic. So, it can be assumed that the rate of a solid state reaction is affected by the degree of crystallinity. Amorphous regions in the crystals pave the way for higher molecular mobility and hence higher reactivity than the crystalline regions. Hence, it has been observed that crystal disorders play a crucial role in enhancing the solid state reactions. Other potent factors influencing the solid state reactions are the increasing surface area and the decreasing surface defects, as solid state reactions are often initiated by defects on the surface of the material. Presence of moisture is ubiquitous and it also plays an influential role in the solid state reactions.

1.7 Homogeneous Kinetics

The concepts of solid state kinetics were developed for homogeneous processes. Whenever a kinetic reaction produces a new phase, either a solid or liquid, there must be growth individually and/or coalescence to bulk of the new phase. Growth of a new phase requires transfer of atoms from the parent to the product lattice causing the interface to advance through the parent crystal. When kinetic reaction occurs at random throughout a parent phase which is chemically homogeneous and free from imperfections, it is called homogeneous kinetics. A simple example of homogeneous kinetics is found in solidification of a pure liquid (Fig. 1.2). In practice, however, even the so-called pure liquids contain many impurity atoms and thus homogeneous reactions are hard to achieve. One way to obviate this problem and study homogeneous kinetics better is to disperse liquid droplets in another liquid of much lower solidification temperature and then cool the whole system gradually.[1] In homogeneous kinetics, kinetic studies were usually directed towards obtaining rate constants that can be used to describe the progress of a reaction. Mechanistic interpretations usually involve identifying reasonable reaction model because information about individual reaction steps is often difficult to obtain.

Fig. 1.2 Schematic representation of **a** homogeneous kinetic mechanism

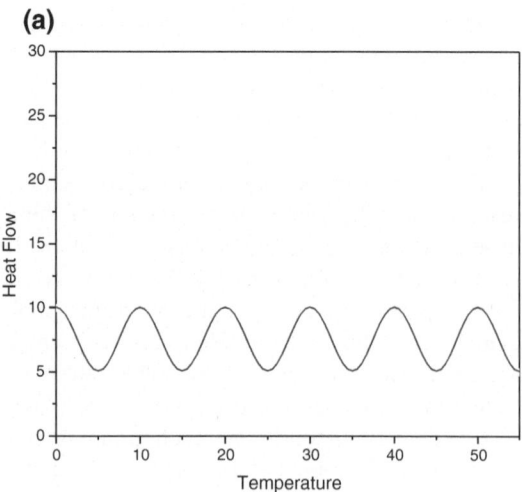

---

[1] Attempts have also been made to study homogeneous kinetics of Co by subjecting levitated Fe–C alloy droplets in an oxidizing gas.

1.8 Heterogenous Kinetics

Recently, considerable progress has been made in the field of homogeneous kinetics and the obtained results are promising for breaking the vicious circle of numerous essential issues that have accumulated around heterogeneous kinetics. Understanding the kinetics of heterogeneous processes is vital to establish product conversion rates and to identify the underlying mechanism of the reaction. A variety of microscopic processes occurring in the core/surface (for example, adsorption, desorption, diffusion, reaction, reconstruction and structure ordering) play essential roles in heterogeneous kinetics. It should be noted that the main aspect in the kinetics of heterogeneous processes is that the total reaction rate does not necessarily have simple mass-action dependence. Therefore, heterogeneous kinetic reactions represent systems far from thermodynamic equilibrium. It can thus be expected that the kinetic equations describing a heterogeneous process are rather involved to formulate and quite difficult to solve (Fig. 1.3).

The conception of nuclei borrowed from biology has had a considerable impact on the theory of heterogeneous processes. This conception has come to chemistry through physics. In the present context, it is worthwhile to mention that (i) a reaction starts from the formation of a stable nuclei leading to the ensemble of growing nuclei, i.e. the reaction zones are multiply connected at the beginning stages; (ii) Subcritical and critical nuclei are small, hence it is extremely difficult to observe them experimentally. Hence, the considerations concerning the mechanism are mainly speculative. In this way, the simultaneous proceeding of various conjugated chemical and physical steps is formalized as the appearance and evolution of the multiply connected reaction zone. In the available literature of heterogeneous processes, the discussion of theoretical aspects is practically never

Fig. 1.3 Schematic representation of heterogeneous kinetic mechanism

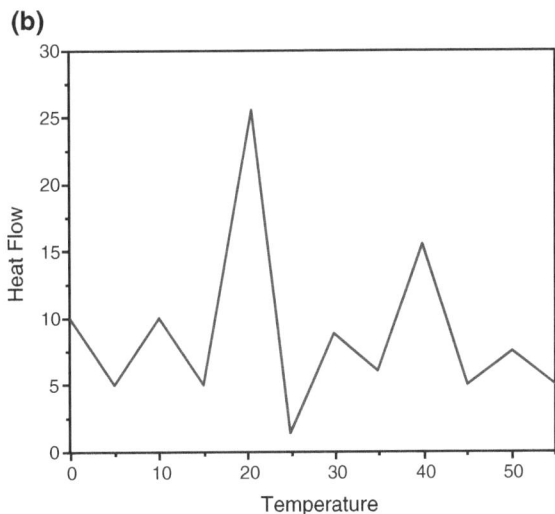

based directly on the work. But when it comes to the questions concerning the essential issues accumulated in heterogeneous kinetics, the development of a realistic and accurate method is required to develop, which will provide a deeper understanding of their roots.

1.9 Controversies in Solid State Kinetic Studies

Generally conventional kinetic studies were confined to isothermal conditions, although it is known that an isothermal reaction is an abstraction and most practical processes involve gradual heating up of the reactants with reactions progressing under rising and fluctuating temperature conditions. Therefore, the kinetic studies conducted under isothermal conditions seldom reflect characteristics of a realistic situation. Accordingly, the estimated isothermal kinetic parameters and reaction rates may not remain valid for actual processes taking place non-isothermally. Considering the above limitations, the development of non-isothermal kinetics becomes an urgent need. But the mathematical approaches for analysing non-isothermal kinetic data suffers from some noted limitations—most of these approaches give incomplete coverage of the subject with cumbersome mathematical procedures and they are limited by a great deal of approximations and assumptions.

Solid state kinetics were developed from the kinetics of homogeneous systems, i.e. liquids and gases. As it is well known, the Arrhenius equation associates the rate constant of a simple one-step reaction with the temperature through the activation energy (E_a) and pre-exponential factor (A). It was assumed that the activation energy (E_a) and frequency factor (A) should remain constant; however this does not happen in the actual case. It has been observed in many solid state-reactions that the activation energy may vary as the reaction progresses which were detected by the isoconversional methods. While this variation appears to be contradictory with basic chemical kinetic principles, in reality, it may not be [15].

However, many reports are available on the true variation of the activation energy, found in both homogeneous and heterogeneous processes. In the solid state, a variation in activation energy could be observed for an elementary reaction due to the heterogeneous nature of the solid sample or due to a complex reaction mechanism.

Variable activation energy during the progress of elementary reactions is attributed to the change in kinetics. This is not usual for homogeneous reactions which occur between freely moving, identical reactant molecules with random collisional encounters that are usually unaffected by product formation. However, reacting entities in a solid sample are not isolated but interact strongly with neighbouring molecules or particles. Therefore, during such a reaction, reactivity may change due to product formation, crystal defect formation, intracrystalline strain or other similar effects. Experimental variables could also affect the solid state reactivity that would change the reaction kinetics by affecting heat or mass transfer at a reaction interface.

When the reaction is affected by two or more elementary each steps having unique activation energy, effecting the rate of product formation, the reaction is complex reaction. In such a reaction, a change in the activation energy as the reaction progresses would be observed. This change depends on the contribution of each elementary step, which gives an effective activation energy that varies with reaction progress.

The temperature dependence of the rate constant is related by the Arrhenius equation. However, there was a lot of controversy surrounding the temperature dependence of rate constants with many workers proposing several forms of rate constant temperature dependency. These equations were empirically derived based on quality of fit. Mere selection of an equation because it gives a reasonable fit to the data is not a sufficient reason for its acceptance, as most of the cited equations will reasonably represent the same experimental data. This occurs because kinetic studies are most often conducted in a narrow temperature range, which makes $1/T$, T and $\ln T$ (i.e. independent variables in these equations) linearly related to one another. As kinetics developed, most of these equations disappeared because they were theoretically unsound. Only the Arrhenius equation survived the test of times as well all the criticisms of various workers in the field of kinetics. But the controversy and confusion surrounding reaction rate temperature dependence still affect researchers in heterogeneous kinetics. Many questions have been raised about the analysis and calculation methods used to study solid-state kinetics. The use of non-isothermal experiments has been criticized in favour of isothermal experiments for two reasons—first, temperature is an experimental variable in non-isothermal analyses while it is fixed in isothermal analyses, which reduces the total number of variables; secondly, kinetic parameters obtained by both isothermal and non-isothermal experiments are usually not in agreement. On the other hand, non-isothermal studies are considered more convenient than isothermal studies because a sample is not subjected to a rapid temperature rise to a reaction temperature (i.e. heat-up time) in which reaction could occur but not be measured, thus introducing errors in the analysis. This is especially true if the isothermal temperature is high because some decomposition probably occurs before the fixed temperature study is initiated.

References

1. Li Y, Na SC, Lu ZP, Feng YP, Lu K (1998) Phil Mag Lett 78:37
2. Mitra A, Palit S, Chattoraj I (1998) Phil Mag B 77:1681
3. Papazian JM (1988) Met Trans A 19A:2945
4. DeIasi R, Adler PN (1977) Met Trans A 8A:1185
5. Li D, Wang X, Xiong G, Lu L, Yang X, Wang X (1997) J Mater Sci Lett 16:493
6. Chen LC, Spaepen F (1988) Nature 336:366
7. Cao X, Kottypin Yu, Kataby G, Prozorov R, Gedanken A (1995) J Mater Res 10:2952

8. Martin TP, Naher U, Bergmann T, Gohlich A, Lange T (1991) Chem Phys Lett 196:113
9. Laidler KJ (1984) J Chem Educ 61:6
10. Khawam A, Flanagan DR (2006) J Phys Chem B 110:17315–17328
11. Vyazovkin S, Wight CA (1999) Thermochim Acta 340–341:53–68
12. Khawam A, Flanagan DR (2005) Thermochim Acta 436:101–112
13. Kissinger HE (1957) Anal Chem 29:1702–1706
14. Paul PC, Curtin DY (1973) Acc Chem Res 6:217–225
15. Khawam A, Flanagan DR (2006) J Pharm Sci 95:3

Chapter 2
Material Development and Process

Recently, there has been a significant increase of interest in fabricating oxide materials that consist of nanosized particles ranging in mean diameter from 1 to 100 nm. The interest in these materials has been stimulated by the fact that, owing to the small size of the building blocks (particles, grain or phase) and the high surface-to-volume ratio, these materials are expected to demonstrate unique mechanical, optical, electronic and magnetic properties. The fabrication of these materials of perfect nanometre-scale crystallites, identically replicated in unlimited quantities in such a state that they can be manipulated and understood as pure macromolecular substances, is an ultimate challenge in modern materials research with outstanding fundamental and potential technological consequences. Interest in magnetic nanoclusters has increased in the past few years by virtue of their novel properties, which have promising broad spectrum applications from biological tagging to recording devices [1–6]. Significantly, γ-Fe$_2$O$_3$ is considered to be the most promising material in this group and is currently finding a variety of applications [7–11]. Therefore, the idea of synthesizing iron oxide particles in their nanocrystalline form for the purpose of superior technological and wider commercial exploitation is gaining a lot of momentum.

Many novel processing techniques have been developed so far to synthesize ultrafine iron oxide powder, e.g. hydrothermal reaction [12], plasma-enhanced chemical vapour deposition [13], sol–gel synthesis [14], spray pyrolysis [15], flame pyrolysis [16] and mechanical activation [17]. However, most of these processes suffer from some serious limitations. First, stringent control over various process parameters is required and secondly, production yield is very low and obviously, not cost effective. In some processes, due to high energy of activation, the resulting powders often exhibit poor particle characteristic represented by a wide particle size distribution and irregular particle morphology, together with a substantial degree of particle aggregation. In addition to the above, most processes require high energy and sophisticated instrumentation. In view of these limitations, the present investigation was primarily designed with an objective to develop a simple and reliable preparation technique, which will involve low temperature treatment, low energy consumption and minimum requirements for sophisticated instrumentation.

P. Deb, *Kinetics of Heterogeneous Solid State Processes*, SpringerBriefs in Materials, DOI: 10.1007/978-81-322-1756-5_2, © The Author(s) 2014

One of the most commonly used techniques to produce monodispersed particles (uniform in size and shape) is precipitate from homogeneous solutions [18], which can be employed for the preparation of uniform colloidal particles. However, almost all studies have been confined to aqueous solutions and little attention has been paid to non-aqueous media. It is to be noted that in aqueous route, a small change in the salt concentration or in the pH over critical regions produces particles of different chemical composition and morphology. Moreover, the growth of the particles cannot be restricted in the aqueous medium during precipitation. In contrast to the above, in non-aqueous route, one may proceed by decomposition of organometallic compounds that will eventually yield uniform particles. In recent times, organic fatty acids, e.g. stearic acid, oleic acid, palmitic acid, etc., are used as the precursors for this purpose [18–20]. Considering these aspects, an attempt has been made here to prepare monodispersed and uniform particles of iron oxide from an organic precursor of stearic acid (Fig. 2.1). This will be followed by the evolution kinetics study of γ-Fe$_2$O$_3$ nanoparticles from the organic precursor during heating.

The synthesis of γ-Fe$_2$O$_3$ nanoparticles is based on the decomposition and subsequent reduction of the intermediate/complex of Fe(O)(stearate) obtained by thermolysis of iron(III) nitrate in a non-aqueous stearic acid medium. The synthesis procedure was as follows. A homogeneous solution was prepared by gradual addition of a calculated amount of Fe(NO$_3$)$_3$. 9H$_2$O to a known amount of molten stearic acid. To have controlled synthesis, the molar ratios of stearic acid to hydrated Fe(III) nitrate was taken in an optimized ratio. The homogeneous solutions prepared with the above composition were thermolyzed separately at 125 °C until evolution of brown fumes of NO$_2$ ceased. At this stage, the solution became viscous. The viscous mass was allowed to cool and solidify in air. The solidified mass so obtained was treated with 80 ml of tetrahydrofuran (THF) from which the precipitates were collected by centrifugation. The precipitated mass was then dried at 70 °C in an air oven for several hours.

The chemical and compositional characteristics of the as-prepared sample were analysed by FTIR spectroscopy analysis. Figure 2.2 represents the FTIR spectrum of the as-prepared powder precipitates. The peaks observed in the spectrum at 692 and 788 cm^{-1} can be assigned to the deformation vibration of Fe–OH groups and the band with the peak at 3285 cm^{-1} is assigned to the O–H stretching vibration of the above group. O–H bending vibration is reflected in the spectrum by the peaks observed at 1042 and 1108 cm^{-1}. The peak observed at 1355 cm^{-1} can be assigned to the characteristic—CH3 bending vibration. Presence of nitro compounds (C–NO$_2$) is indicated by the band observed between 1475 and 1702 cm^{-1}. The peaks observed at 2835 and 2904 cm^{-1} can be assigned to C–H stretching vibration. The above spectroscopic observations suggest that the as-prepared powder sample consist of residual stearic acid, nitro compounds and FeOOH in an intermediate/complex of Fe(O)(stearate).

The surface functionalities of the as-prepared sample have been investigated by XPS characterization. Figure 2.3 presents the C1s and O1s XPS spectra for the sample. This stearic acid coated sample shows a rather complex composition of the C1s peak, where at least four different components are present. Aliphatic carbon

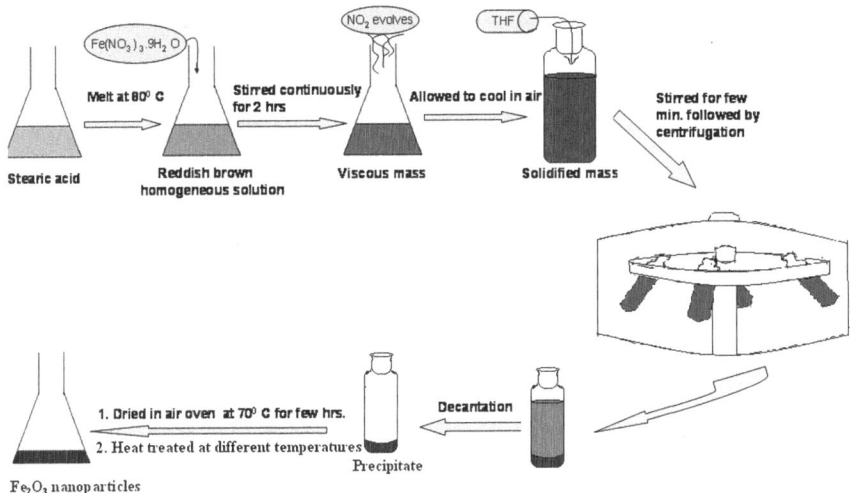

Fig. 2.1 Schematic diagram of synthesis of iron oxide nanoparticles

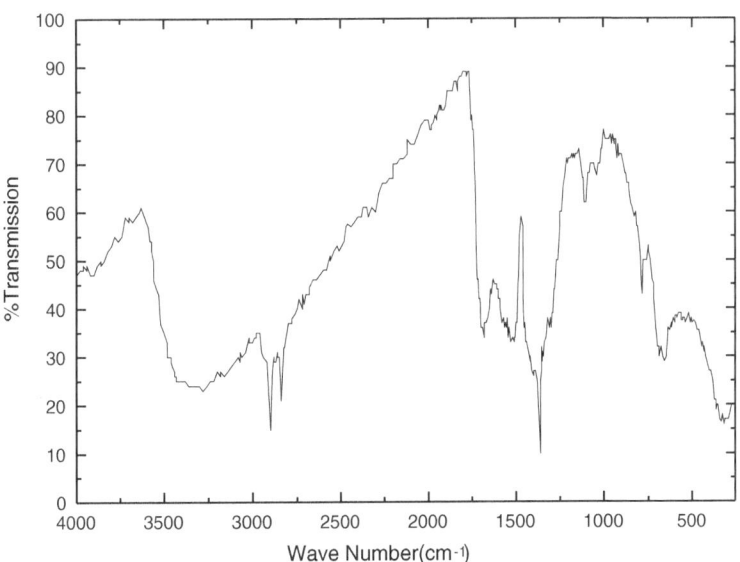

Fig. 2.2 FTIR spectrum of the as prepared powder precipitates

from stearic acid is present at 285 eV [21]. Two peaks at higher binding energies, 287 and 290.2 eV, are for the carbon atoms in the vicinity of oxygen which indicates to the C atoms of carboxyl group, respectively [22]. In addition, one peak at 283.2 eV can be explained by carbon in contact with an electropositive species which points towards the probability of formation of carbide or something else

Fig. 2.3 High resolution XP spectra of as prepared sample on Si wafer in **a** C1s and **b** O1s regions

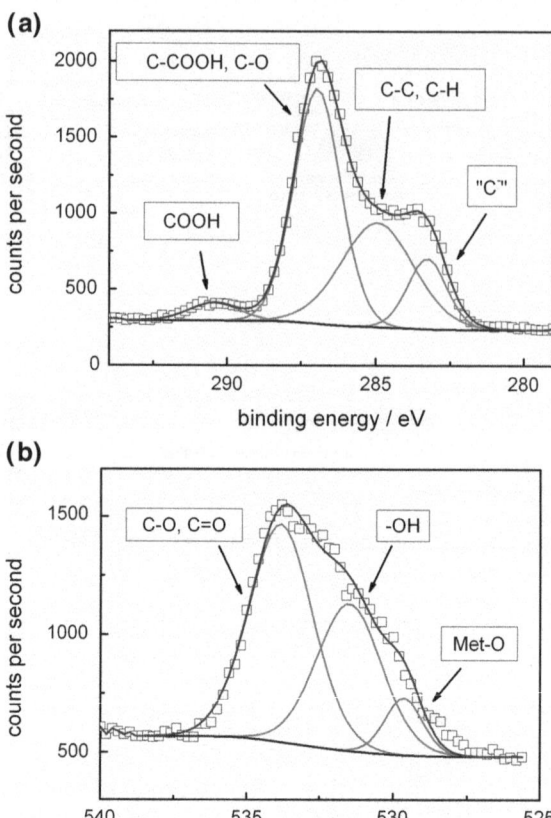

which needs more careful observation [23]. The O1s region also contains several components; three are easily distinguishable; among them the peak at 533.8 eV is due to oxygen bound to carbon [22]. Another peak at 531.5 eV is due to the presence of −OH groups which may come from the iron oxyhydroxides (FeOOH) formed during synthesis [24]. The peak at 529.7 eV is due to iron oxide [25]. It is worthwhile to mention here that survey scans show almost no signal of Fe. This means that there is a thick shell (>5 nm) of organic material around the particles.

The XPS investigation into the surface characteristics of the sample with stearic acid as surfactant concludes that the resulting particles have a thick shell (>5 nm) of organic material around the inorganic part, which implies that the stearic acid coating consists of more than one layer. Presence of strongly and weakly bound stearic acid layers over sample surface is evidenced.

Nanosized particles of γ-Fe_2O_3 were obtained by non-isothermally heating the precipitates from room temperature to 300 °C. The desired temperature of the furnace was controlled by a proportional integral and derivative (PID) controller with an accuracy of ±1 °C.

References

1. Skumryev V, Stoyanov S, Zhang Y et al (2003) Nature 423:850
2. Sarikaya M, Tamerler C, Jen A et al (2003) Nat Mater 2:577
3. Sun S, Murray CB, Weller D et al (2000) Science 287:1989
4. Niemeyer CM (2001) Angrew Chem Int Ed Engl 40:4128
5. Thompson DA, Best JS (2000) IBM J Res Dev 44:311
6. Kodama RH (1999) J Magn Magn Mater 200:359
7. Buriak JM (2004) Nat Mater 3:847
8. Pankhurst Q, Connolly J, Jones SK et al (2003) J Phys D Appl Phys 36:R167
9. Ngo AT, Pileni MP (2001) J Phys Chem B 105:53
10. Hyeon T, Lee SS, Park J et al (2001) J Am Chem Soc 123:12798
11. Sugimoto M (1999) J Am Ceram Soc 82:269
12. Sahu KK, Rath C, Mishra NC, Anadn S, Das RP (1997) J Coll Interf Sci 185:402
13. Liu Y, Zhu W, Tan OK, Shen Y (1997) Mater. Sci. Engg. B47:171
14. Sugimoto T, Sakada K (1992) J Coll Interf Sci 152:587
15. Morales MP, de.Julian C, Gonzales JM, and Serna CJ (1994) J Mater Res 9 135
16. Grimm S, Schultz M, Barth S, Miller R (1997) J Mater Sci 32:1083
17. Liu X, Ding J, Wang J (1999) J Mater Sci 14:3355
18. Matijevic E (1993) Chem Mater 5:412
19. Cheng FX, Jia JT, Xu ZG, Zhou B, Liao CS, Yan CH (1999) J Appl Phys 86:2727
20. Shafi KVPM, Gedanken A, Prozorov R, Balogh J (1998) Chem Mater 10:3445
21. Beamson G, Briggs D (1992) High resolution XPS of organic polymers: the scienta ESCA 300 database. Wiley, New York
22. Dilks A (1981) J. Polym Sci Polym Chem Ed 19:1319. Standard reference data. http://www.nist.gov/srd/
23. Goretzki H, Rosenstiel PV, Mandziej S, Fres. Anal.Z (1989) Chem. 333:451. Standard reference data. http://www.nist.gov/srd/
24. McIntyre NS, Zetaruk DG (1977) Anal Chem 49:1521
25. Haber J, Stoch J, Ungier L (1976) J Electron Spectrosc Relat Phenom 9:459

Chapter 3
Non-isothermal Kinetic Analysis Method

Solid state kinetics lays its foundation on the basis of the experiments that are carried out in absolute isothermal conditions. However, although it is known that thermally activated processes involve gradual heating up of the reactants and the reactions progress under rising and fluctuating temperature conditions, conventional kinetic studies were confined to only isothermal conditions due to limitation of performing. In addition, the mathematical procedures developed for analysing non-isothermal kinetic data contain numerous approximations, assumptions and controversies [1–5]. These procedures fail to reflect the characteristics of a realistic situation. In addition to this, a priori knowledge of the reaction mechanism is required. The reaction mechanism is either assumed or identified by carrying out isothermal kinetic experiments. The methods [6–9] most widely used to evaluate activation energy from the non-isothermal calorimetric data are also based on a particular type of assumed mechanism. Nonetheless, characteristic reaction mechanism(s) do exist for all the reactions occurring under non-isothermal conditions. Therefore, the development of a more reliable and accurate non-isothermal kinetic analysis method is long overdue. In the present study, it has been shown that the actual reaction mechanism under non-isothermal conditions can be directly identified unambiguously through a newly developed technique.

3.1 Identification of the Kinetic Law

The analysis is based on the fact that the general kinetic equation for solid state processes is generally represented by

$$g(\alpha) = k(T)t \tag{3.1}$$

where $g(\alpha)$ is the reaction mechanism and is expressed as an appropriate function of fractional reaction (α), $k(T)$ is the specific rate constant at a specified temperature T and t is the time [10]. Here, the fractional conversion (α) values are required to be recorded against rising temperature T at different heating rates. The variation in

P. Deb, *Kinetics of Heterogeneous Solid State Processes*, SpringerBriefs in Materials, DOI: 10.1007/978-81-322-1756-5_3, © The Author(s) 2014

fractional conversion as a function of temperature at specified heating rate β_i can be expressed conveniently by a relationship of the type

$$\alpha_{\beta i} = \varphi_{\beta i}(T) \tag{3.2}$$

where $\alpha_{\beta i}$ is the fractional conversion at a heating rate β_i at temperature T and $\varphi_{\beta i}$ (T) is an appropriate function of temperature and is the heating rate identity. Therefore, at any intermediate temperature within the limits of experimental temperature range $\alpha_{\beta i}$ can be easily evaluated by using spline interpolation technique [11].

During the non-isothermal experimentation, a linear rise in temperature can be represented by

$$T_f = T_i + \beta_i t_{\beta i} \tag{3.3}$$

where T_i is the starting temperature of the reaction corresponding to the heating rate β_i and $t_{\beta i}$ is the time spent in attaining the desired temperature T_f.

Therefore,

$$t_{\beta i} = (T_f - T_i)/\beta_i. \tag{3.4}$$

Now, considering any one of the experimental heating rate β_n, relationship of the type

$$\beta_n(T_f - T_i)/\beta_i(T_f - T_n) = t_{\beta i}/t_{\beta n} = \theta \tag{3.5}$$

can be formed, where θ is the dimensionless time and T_i and T_n are the onset temperatures of the reaction corresponding to the heating rates β_i and β_n. $t_{\beta i}$ and $t_{\beta n}$ are the time taken to raise the temperature from T_i or T_n to T_f.

Considering a specified temperature T_f, which is common to both the heating rates β_i and β_n, the solid state reaction mechanism which remains operative, can be identified by using the expression shown below.

$$g\{\varphi_{\beta i}(T_f)\}/g\{\varphi_{\beta n}(T_f)\} = t_{\beta i}/t_{\beta n} = [\beta_n(T_f - T_i)]/[\beta_i(T_f - T_n)] = \theta \quad (3.6)$$

The above expression is formed by combining Eqs. (3.1), (3.2), (3.3) and (3.5). The expression is dimensionless and independent of the nature of the system, temperature, heating rate or any other factors that influence the rate. Therefore, it is evident that every possible reaction mechanism will exhibit dimensionless characteristic values at particular temperatures for the heating rates β_i and β_n under consideration. Hence, it is quite possible that the actual mechanism of a solid state reaction under non-isothermal condition can be identified by matching the characteristic dimensionless theoretical values of $g\{\varphi_{\beta i}(T_f)\}/g\{\varphi_{\beta n}(T_f)\}$ for different mechanisms (some of them are listed in Table 3.1) with the experimentally obtained values of $\theta = [\beta_n(T_f-T_i)]/[\beta_i(T_f-T_n)]$.

Table 3.1 Models for different solid state processes

Symbol	Name	Form of $g(\alpha)$
D1	Parabolic	α^2
D2	Valensi Barrer	$\alpha + (1 - \alpha)\ln(1 - \alpha)$
D3	Ginstling Brounsthein	$1 - 2/3\alpha - (1 - \alpha)^{2/3}$
D4	Jander	$\left[1 - (1 - \alpha)^{1/3}\right]^2$
D5	Anti Jander	$\left[(1 - \alpha)^{1/3} - 1\right]^2$
D6	Zhuralev, Lesokhin and Templeman	$\left[(1 - \alpha)^{-1/3} - 1\right]^2$
CG1	Linear Growth	α
CG2	Cylindrical Symmetry	$\left[1 - (1 - \alpha)^{1/2}\right]$
CG3	Spherical Symmetry	$\left[1 - (1 - \alpha)^{1/3}\right]$
NG1	Avrami Erofeev, $n = 1.5$	$[-\ln(1 - \alpha)]^{1/1.5}$
NG2	Avrami Erofeev, $n = 2.0$	$[-\ln(1 - \alpha)]^{1/2.0}$
NG3	Avrami Erofeev, $n = 3.0$	$[-\ln(1 - \alpha)]^{1/3.0}$
NG4	Avrami Erofeev, $n = 4.0$	$[-\ln(1 - \alpha)]^{1/4.0}$
P1	Mempel Power Law	$\alpha^{1/2}$
P2	Mempel Power Law	$\alpha^{1/3}$
P3	Mempel Power Law	$\alpha^{1/4}$
R1	First order reaction	$[-\ln(1 - \alpha)]$
R2	One and a half order reaction	$\left[(1 - \alpha)^{-1/2} - 1\right]$
R3	Second order reaction	$\left[(1 - \alpha)^{-1} - 1\right]$

3.2 Evaluation of Activation Energy

Once the actual reaction mechanism is identified, the activation energy of the reaction can be calculated by taking recourse to a relationship which is similar to that used by Medak [12], Judd [13] and Ozawa [14].

Considering Eq. (2.1) and differentiating w.r.t. time at a particular temperature (T), we get

$$(d\alpha/dt)_T = k(T)/[dg(\alpha)/d\alpha] = k(T)f(\alpha) \qquad (3.7)$$

where $1/[dg(\alpha)/d\alpha] = f(\alpha)$.

However, a lot of controversies were raised [9, 15–17] regarding the actual meaning of the term $d\alpha/dt$. Since then it has been assumed that under non-isothermal conditions $\alpha = f(t, T)$. Hence,

$$d\alpha = (\partial\alpha/\partial t)_T \, dt + (\partial\alpha/\partial T)_t dT. \qquad (3.8)$$

However, the physical interpretation of $(\partial\alpha/\partial T)_t$ is not possible because temperature cannot be varied by keeping the time constant. Moreover, if we assume that the instantaneous change is possible, α cannot change instantaneously. Therefore, mathematical inconsistency will arise with the assumption that α is a state function of t and T. Hence, it is worthwhile to mention that in reality α is a path function [18].

From Eq. (3.3) we have

$$dT_f = \beta_i dt_{\beta i} \quad i.e. \, dt_{\beta i} = dT_f/\beta_i. \tag{3.9}$$

Therefore, by combining Eqs. (3.7) and (3.9) we can write

$$d\alpha/dT_f = k(T)f(\alpha)/\beta_i. \tag{3.10}$$

The temperature dependency of the rate constant $k(T)$ is expressed as [19]

$$k(T) = AT^m \exp(-E/RT). \tag{3.11}$$

However, since the exponential term is much more temperature sensitive than the T^m term, the variation of $k(T)$ caused by the latter is effectively masked. This can be shown in the following way.

Taking logarithms of Eq. (3.11) and differentiating w.r.t. T, we get

$$d[\ln k(T)]/dT = m/T + E/RT^2 = (mRT + E)/RT^2. \tag{3.12}$$

Since $mRT \ll E$ for most solid state reactions, we can ignore mRT and write

$$d[\ln(k)]/dT = E/RT^2 \quad i.e. \quad k = A \exp(-E/RT). \tag{3.13}$$

Substituting, $k = A \exp(-E/RT)$, Eq. (2.10) can be written as

$$d\alpha/dT_f = A \exp(-E/RT)f(\alpha)/\beta_i. \tag{3.14}$$

Rearranging and integrating we get

$$\int\limits_0^\alpha d\alpha/f(\alpha) = (A/\beta_i) \int\limits_0^{T_f} \exp(-E/RT_f)dT_f \tag{3.15}$$

and hence,

$$g(\alpha) \approx ART_f^2 \exp(-E/RT_f)[1 - 2RT_f/E]/\beta_i E$$

$$i.e. \quad g(\alpha)/T_f^2 = AR \exp(-E/RT_f)\left[1 - 2RT_f/E\right]/\beta_i E. \tag{3.16}$$

Taking logarithm on both sides

$$\ln\left[g(\alpha)/T_f^2\right] = \ln\{AR[1 - 2RT_f/E]/\beta_i E\} - E/RT_f. \tag{3.17}$$

The logarithmic term in the RHS of the above equation remains almost constant for close values of β_i. Therefore, a plot of $\ln[g(\alpha)/T_f^2]$ versus $1/T_f$ should yield a straight line with slope E/R, which enables one to calculate the activation energy (E) of the reaction from the slope.

However, manual computation of the theoretical ratios for all the possible mechanisms (refer Table 3.1) and matching them with experimentally obtained values of θ for different temperature and heating rates is laborious. This task has been accomplished by developing a computer program according to the procedure described above. The developed program (Kinlac) identifies the reaction mechanism for a particular fixed temperature for all the heating rates. The program also computes the $\ln[g(\alpha)/T_f^2]$ values and the corresponding $1/T_f$ values required for plotting and computing the activation energy of the reaction.

3.3 Kinetic Equation Under Rising Temperature Condition

The theoretical treatment of kinetic data obtained under rising temperature conditions rests largely on a combination of three equations:

The first is the kinetic law written in the differential form

$$d\alpha/dt = k(T)f(\alpha)\Phi(\alpha, T) \tag{3.18}$$

where $k(T)$ is the temperature-dependent specific rate constant, f and Φ denote functions, t is the time and T is the temperature.

Generally $\Phi(\alpha,T)$ is assumed to be unity, thus

$$d\alpha/dt = k(T)f(\alpha). \tag{3.19}$$

The variation in fractional conversion is a function of temperature at specified heating rate

$$\alpha_{\beta i} = \Phi_{\beta i}(T) \tag{3.20}$$

where $\alpha_{\beta i}$ is the fractional conversion at a heating rate β_i at temperature T and $\Phi_{\beta i}(T)$ is an appropriate function of temperature and is heating rate identity.

The second equation we need is the law describing the temperature co-efficient of the rate constant

$$k(T) = AT^m e^{-E/RT}. \tag{3.21}$$

However, since the exponential term is much more temperature sensitive than the T^m term, the variation of $k(T)$ caused by the latter is effectively masked. This can be shown in the following way:

Taking logarithms of Eq. (3.4) and differential w.r.t. T, we get

$$d[\ln(T)]/dT = m/T + E/RT^2 = (mRT + E)/RT^2. \quad (3.22)$$

Since $mRT \ll E$ for most of the solid state reactions, we can ignore mRT and write

$$d[\ln(k)]/dT = E/RT^2$$

$$k = A \exp\left(-E/RT^2\right). \quad (3.23)$$

In the above equations E is the activation energy, A is the pre-exponential factor and R is the gas constant, m is a constant.

The third equation we need should describe the variation of the temperature T with time. For a linear rate to rise,

$$T_f = T_i + \beta_i t_{\beta i} \quad (3.24)$$

where T_i is the starting temperature of the reaction corresponding to the heating rate β_i and $t_{\beta i}$ is the time spent in attaining the desired temperature T_f. Therefore,

$$t_{\beta i} = \left(T_f - T_i\right)/\beta_i. \quad (3.25)$$

Now considering any one of the heating rates β_n, relationship of the type

$$\beta_n\left(T_f - T_i\right)/\beta_i\left(T_f - T_n\right) = t_{\beta i}/t_{\beta n} = \theta \quad (3.26)$$

can be formed, where θ is the dimensionless time and T_i and T_n are the onset temperatures of the reaction corresponding to the heating rates β_i and β_n. $t_{\beta i}$ and $t_{\beta n}$ are the time taken to raise the temperature from T_i or T_n to T_f.

Corresponding to a specified temperature T_f which is common to the heating rates β_i and β_n, the solid state reaction mechanism which remains operative can be identified by using the expression shown below:

$$g\left\{\Phi_{\beta i}\left(T_f\right)\right\}/g\left\{\Phi_{\beta i}\left(T_f\right)\right\} = t_{\beta i}/t_{\beta n} = \beta_n\left(T_f - T_i\right)/\beta_i\left(T_f - T_n\right) = \theta. \quad (3.27)$$

The above expression is formed by combining Eqs. (3.1), (3.3), (3.7) and (3.9). The expression is dimensionless and independent of the nature of the system, temperature, heating rate or any other factors that influence the rate. Therefore, it is evident that every possible reaction mechanism will exhibit dimensionless characteristic values at a particular temperature for the heating rates β_i and β_n under consideration. Hence, it is quite possible that the actual mechanism of a solid state reaction under non-isothermal condition can be identified by matching the characteristic dimensionless theoretical values of $g\{\Phi_{\beta i}(T_f)\}/g\{\Phi_{\beta i}(T_f)\}$ for different mechanisms with the experimentally obtained values of $\theta = \beta_n(T_f - T_i)/\beta_i(T_f - T_n)$.

It has been assumed that non-isothermal conditions, $\alpha = f(\beta_i, T_f)$. Hence,

$$d\alpha/dt = (\partial\alpha/\partial\beta_i)_{Tf} d\beta_i/dt + (\partial\alpha/\partial T_f)_{\beta_i} dT/dt. \tag{3.28}$$

We have from Eq. (3.7)

$$\beta_i = (T_f - T_i)/t$$

So,

$$t = (T_f - T_i)/\beta_i.$$

Now differentiating

$$dt = [\beta_i d(T_f - T_i) - (T_f - T_i)d\beta_i]/\beta_i^2$$
$$dt = [-(T_f - T_i)d\beta_i]/\beta_i^2. \tag{3.29}$$

Now, Eqs. (2.1) and (2.12) imply that

$$d\alpha/dt = k(T_f)f(\alpha)$$
$$d\alpha/[-(T_f - T_i)d\beta_i]/\beta_i^2 = k(T_f)f(\alpha) \tag{3.30}$$
$$d\alpha/f(\alpha) = [-(T_f - T_i)k(T_f)]/\beta_i^2 d\beta_i.$$

Integrating both sides,

$$\int d\alpha/f(\alpha) = \int [-(T_f - T_i)d\beta_i]/\beta_i^2 d\beta_i$$

$$g(\alpha) = -(T_f - T_i)\left[k(T_f)\int d\beta_i/\beta_i^2 + \int dk(T_f)/d\beta_i \int d\beta_i/\beta_i^2\right]$$

$$g(\alpha) = (T_f - T_i)k(T_f)/\beta_i + (T_f - T_i)\int A\exp(-E/RT_f)(-E/R)\left(-1/T_f^2\right)dT_f/d\beta_i(-1/\beta_i)d\beta_i$$

$$so,\ g(\alpha) = (T_f - T_i)k(T_f)/\beta_i - (T_f - T_i)(EA/R)\int \exp(-E/RT_f)/\left(T_f^2\beta_i\right)dT_f.$$

$$\tag{3.31}$$

Now, we have from Eq. (2.14),

$$\int \left[\exp(-E/RT_f) / (T_f^2 \beta_i) \right] dt_f$$

$$= 1/\beta_i \exp(-E/RT_f) \int \left(1/T_f^2 \right) dT_f - 1/\beta_i \int d/dT_f \left\{ \exp(-E/RT)_f \right\} \int \left(1/T_f^2 \right) dT_f dT_f$$

$$= - \exp(-E/RT_f)/(\beta_i T_f) + E/R\beta_i \int \left[\exp(-E/RT_f)/T_f^2 \right] dT_f$$

$$= - \exp(-E/RT_f)/(\beta_i T_f) + E \exp(-E/RT_f)/(R\beta_i) \int \left(1/T_f^2 \right) dT_f$$

$$\quad - E/R\beta_i \int d/dT_f \exp(-E/RT_f)(-E/R)\left(-1/T_f^2 \right) \int \left(1/T_f^3 \right) dT_f$$

$$= - \exp(-E/RT_f)/(\beta_i T_f) - E \exp(-E/RT_f)/2RT_f^2 \beta_i + E^2/2R^2 \beta_i \int \exp\left(-E/RT_f^4 \right) dT_f$$

$$= - \exp(-E/RT_f)/\beta_i \left[1/T_f + E/2RT_f^2 + E/3RT_f^3 + \ldots \right].$$

Therefore, Eq. (2.14) gives

$$g(\alpha) = (T_f - T_i) k(T_f)/\beta_i - (T_f - T_i) EA \exp(-E/RT_f)/R\beta_i \left[-1/T_f - E/2RT_f^2 - E/3RT_f^3 - \ldots \right]$$

$$= (T_f - T_i) k(T_f)/\beta_i - (T_f - T_i) k(T_f)/\beta_i \ln[1 - E/RT_f]$$

$$g(\alpha)/T_f^2 = k(T_f)/\beta_i \left[(T_f - T_i)/T_f^2 - (T_f - T_i)/T_f^2 \ln\left[1 - E/RT_f \right] \right]$$

So, $\ln\left[g(\alpha)/T_f^2 \right] = \ln\left[A(T_f - T_i)/T_f^2 - A(T_f - T_i)/T_f^2 \ln\{1 - E/RT_f\} \right] - E/RT_f - \ln \beta_i$

$$= \ln\left[A(T_f - T_i)/T_f^2 - A(T_f - T_i)/T_f^2 \ln\{1 - E/RT_f\} \right]$$

$$\quad - E/RT_f \left[1 + RT_f/E \ln \beta_i \right]$$

$$(3.32)$$

The logarithmic term on the right-hand side of the above equation remains almost constant. Therefore, a plot of $\ln[g(\alpha)/T_f^2]$ versus $1/T_f$ should yield a straight line with slope $E/R[1 + RT_f/E \ln\beta_i]$, which depends upon heating rate β_i. Since (RT_f/E) is very small, hence dependence of slope is on the term $\ln\beta_i$.

References

1. Coats AW, Redfern JP (1964) Nature 201:68
2. Doyle CD (1965) Nature 207:290
3. MacCullum JR, Tanner J (1970) Nature 225:1127
4. Gilles JM, Tompa H (1971) Nat Phys Sci 229:57
5. MacCullum JR, Tanner J (1971) Nat Phys Sci 232:41
6. Papazian JM (1988) Met Trans A 19A:2945
7. DeIasi R, Adler PN (1977) Met Trans A 8A:1185

8. Chen LC, Spaepen F (1988) Nature 336:366
9. Kissinger HE (1957) Anal Chem 29:1702–1706
10. Ray HS (1993) Kinetics of metallurgical reactions. Oxford and IBH Publishing Co. Pvt. Ltd., New Delhi, p 212
11. Skumryev V, Stoyanov S, Zhang Y et al (2003) Nature 423:850
12. Medek J (1976) J Thermal Anal 10:211
13. Judd MD, Pope MI (1973) J. Thermal Anal 5:501
14. Ozawa T (1973) J Thermal Anal 5:499
15. Vyazovkin S, Wight CA (1999) Thermochim Acta 340–341:53–68
16. Thompson DA, Best JS (2000) IBM J Res Dev 44:311
17. Kodama RH (1999) J Magn Magn Mater 200:359
18. Felder RM, Stahel EP (1970) Nature 228:1085
19. Paul PC, Curtin DY (1973) Acc Chem Res 6:217–225

Chapter 4
Kinetics of a Solid State Process

The kinetic study of a solid state process extends several significant information about the process. A critical study of the kinetics gives a better understanding of the process as well as also helps in predetermining the equilibrium state. Therefore, in this chapter I have tried to analyze the kinetic data on the evolution of iron oxide nanoparticles to yield a better insight on the reaction mechanism and process parameters. Moreover, the study will help one to design and control the process parameters more accurately and effectively.

The applicability and validity of the method mentioned in the previous chapter was tested by performing DSC experiments over powdery samples of Fe(O)(stearate) intermediate/complex, processed from a nonaqueous organic precursor medium [1, 2]. The DSC experiments were carried out in static air medium at three different heating rates, 6, 8 and 10 °C/min respectively. Heating of the powdery sample liberates γ-Fe$_2$O$_3$ nanoparticles by decomposition of the intermediate/complex of Fe(O)(stearate). The kinetic law, which remains operative during the liberation of γ-Fe$_2$O$_3$ nanoparticle, is unambiguously identified. The relevance of the kinetic law in the present context is established by correlating and studying the microstructure of the nanoparticles.

4.1 Kinetic Analysis of a Solid State Reaction

The DSC plots (refer Fig. 4.1) of the samples for the three heating rates exhibit similar thermal characteristics with sharp exothermic decomposition/oxidation peak of the powdery precipitates above 175 °C. The peak temperatures for the heating rates 6, 8 and 10 °C/min are observed at 256.0, 260.9 and 264.7 °C respectively. The fractional conversion (α) values were evaluated from the DSC plots by measuring the relative area under the curve with respect to the temperature and are shown in Fig. 4.2.

Kinetic analysis based on the procedure described in previous chapter reveals that the kinetic law, which remains operative during the evolution of γ-Fe$_2$O$_3$ nanoparticle from Fe(O)(stearate) intermediate/complex, is of the form

P. Deb, *Kinetics of Heterogeneous Solid State Processes*, SpringerBriefs in Materials, DOI: 10.1007/978-81-322-1756-5_4, © The Author(s) 2014

Fig. 4.1 DSC plots for the powdery samples of Fe(O)(stearate) intermediate/complex at different heating rates

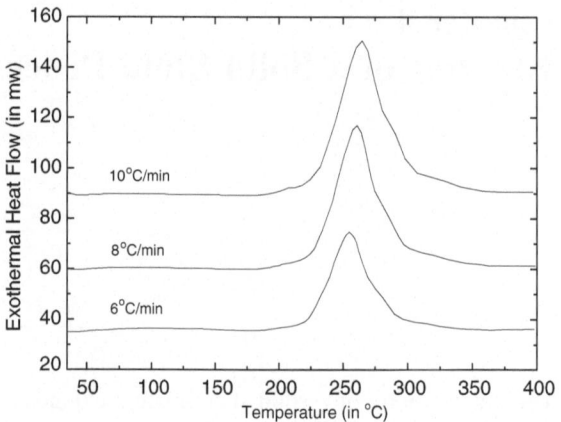

Fig. 4.2 Effect of heating rates on the kinetics of Fe(O)(stearate) intermediate/complex powder samples

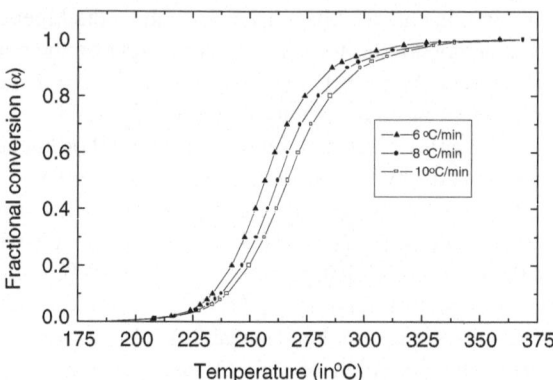

$[-\ln(1 - \alpha)]^{1/2.0} = kt$. This kinetic law represents the solid state reaction model for nucleation and growth type of mechanism [3]. It suggests that (i) γ-Fe_2O_3 particles will be nucleated and precipitated out from the Fe(O)(stearate) intermediate/complex during heating and (ii) the growth of the fine nuclei of γ-Fe_2O_3 particles will be restricted to the surface only (two dimensional growth).

Figure 4.3 represents the $\ln[g(\alpha)/T^2]$ versus $1/T$ plots for evaluating the activation energy. The activation energy as computed from the slope of the lines of Fig. 4.3 has been found to be 115 kJ/mole.

This $\ln[g(\alpha)/T^2]$ versus $1/T$ plots (refer Fig. 4.3) for evaluating the activation energy has been compared with the widely used Kissinger plot [4, 5] (refer Fig. 4.4). The activation energy computed from the Kissinger plot is 130 kJ/mole. The difference in the evaluated activation energy values can be explained on the basis of the fact that irrespective of the reaction order, Kissinger relationship,

$$\{d[\ln(\phi/T_m^2)]\}/\{d(1/T_m)\} = -E/R \qquad (4.1)$$

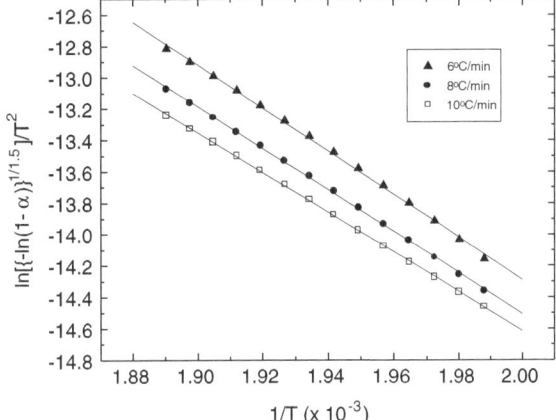

Fig. 4.3 $\text{Ln}[g(\alpha)/T^2]$ versus $1/T$ plots for different heating rates. The activation energy of the reaction, evaluated from the slope of the *lines*, is 115 kJ/mole

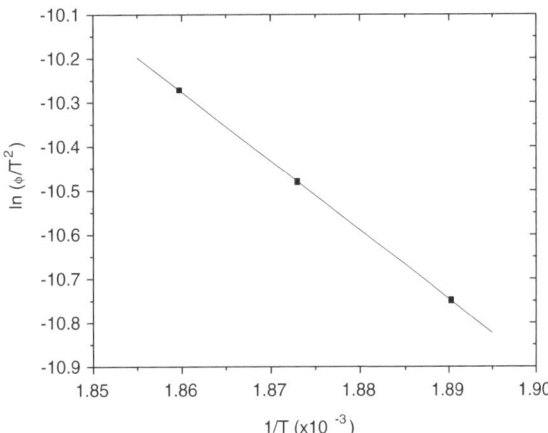

Fig. 4.4 Kissinger plot to estimate the activation energy of the reaction. The activation energy of the reaction evaluated from the slope of the *line* is 130 kJ/mole

where T_m the peak temperature of the transformation at a heating rate φ and R the gas constant, has been derived from an assumed mechanism of the type

$$1/(n-1)\left\{1\big/(1-\alpha)^{n-1}-1\right\} = kt.$$

In comparison to the above, the present method shows that the reaction mechanism can be identified unambiguously by analyzing the non-isothermal kinetic data. Therefore, the computation of activation energy based on this identified mechanism will be more realistic and accurate.

The TEM studies of the samples obtained after heating the sample from room temperature to 350 °C at a rate of 8 °C/min were also carried out. It was evident from the micrographs that the evoluted particles are very fine (average size ∼ 10 nm) possessing two-dimensional morphological features. The presence of this feature corroborates and upholds the validity of the identified kinetic law.

References

1. Deb P, Basumallick A (2001) J Mater Res 16:3471
2. Deb P, Basumallick A, Chatterjee P, Sengupta SP (2001) Scripta Mater 45:341
3. Ray HS (1993) Kinetics of metallurgical reactions. Oxford and IBH Publishing Co Pvt Ltd, New Delhi, p 212
4. Kissinger HE (1957) Anal Chem 29:1702–1706
5. Mitra A, Palit S, Chattoraj I (1998) Phil Mag B 77:1681

Chapter 5
Kinetics of the Heterogeneous Solid State Process

Heterogeneous solid state processes do not involve the physical parameters that characterize the state of reaction system homogeneously during the reaction. A close examination of the DSC plots (refer Fig. 4.1) of the samples at different heating rate reveals the asymmetric nature, which implies the occurrence of intermediate reactions [1, 2] during the process. Here, it is felt that identification of these intermediate reactions and the kinetic laws, which remain operative, will help one to cast a deeper insight into the evolution of iron oxide nanoparticles from the organic precursor.

Figures 5.1, 5.2, 5.3 show the deconvoluted patterns of the DSC curves of the sample at the heating rates 6, 8 and 10 °C/min, respectively. In order to identify the intermediate reactions and the kinetic laws, the temperature range related to the main exothermic peak was subjected to a carefully conducted deconvolution exercise. The deconvolution was preceded by plotting d^2H/dT^2 as a function of T (H being the heat flow) within the said temperature interval for identifying the number of peaks. Deconvolution of the DSC patterns was done by keeping the area under the main curve constant and assuming the intermediate reaction thermal profiles to be symmetrical.

On the basis of the above-mentioned procedure, the main DSC pattern has been deconvoluted into five segments and the analysis of the deconvoluted curves as described below suggests the following:

(i) The first curve exhibiting the small exothermic peak which begins at 188 °C and ends at 224 °C arises probably due to the partial decomposition/oxidation of Fe(O) (stearate) intermediate/complex into simpler compounds. Here, it would be worthwhile to mention that although decomposition reactions are endothermic in nature, the net heat liberated remains positive due to the predominance of oxidation/combustion reactions.

(ii) The second deconvoluted exothermic curve, which begins at 200 °C and ends at 317 °C, represents the combustion/oxidation of attached organic matter with the liberation of iron oxide nanoparticles.

P. Deb, *Kinetics of Heterogeneous Solid State Processes*, SpringerBriefs in Materials, DOI: 10.1007/978-81-322-1756-5_5, © The Author(s) 2014

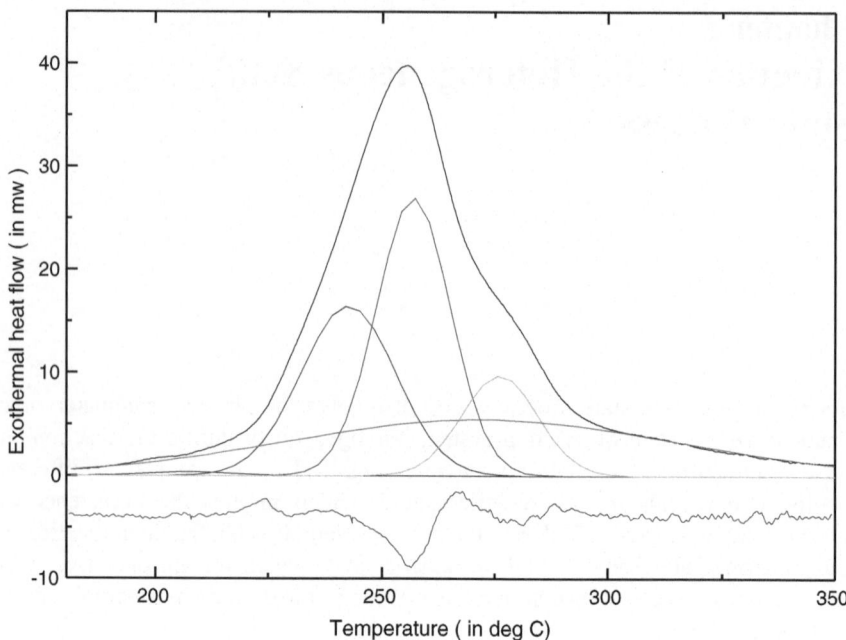

Fig. 5.1 Deconvoluted patterns for the DSC curve of the sample with 6 °C/min heating rate

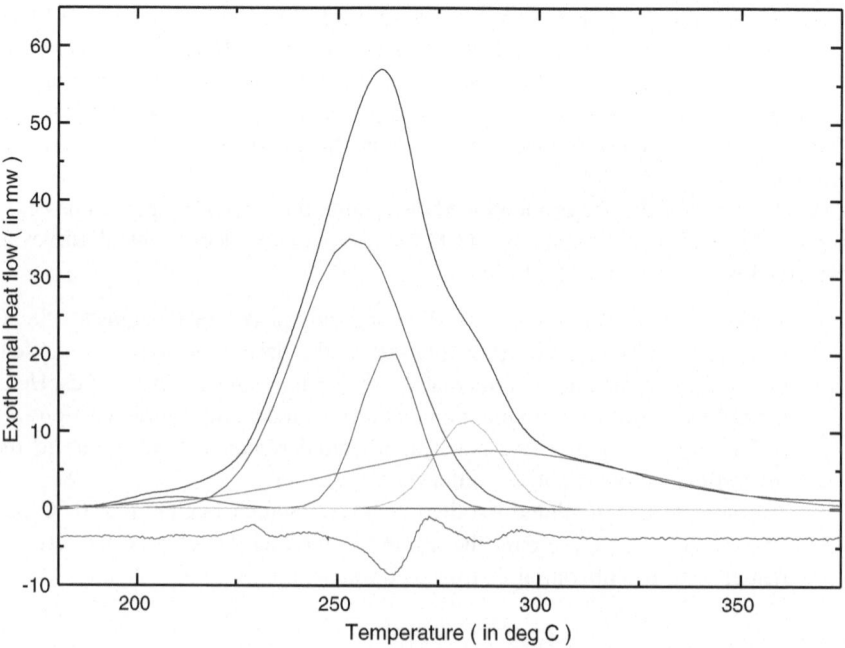

Fig. 5.2 Deconvoluted patterns for the DSC curve of the sample with 8 °C/min heating rate

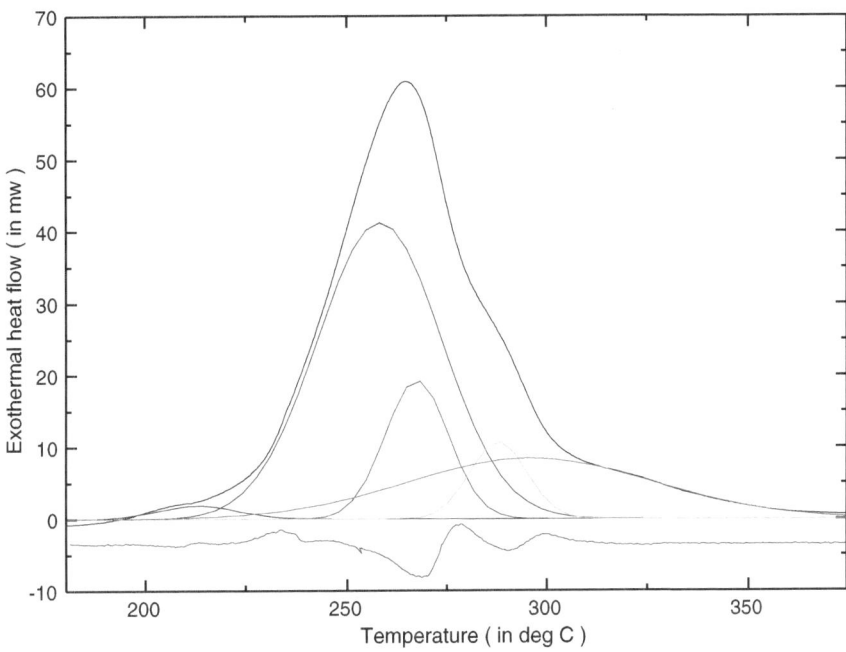

Fig. 5.3 Deconvoluted patterns for the DSC curve of the sample with 10 °C/min heating rate

(iii) The third deconvoluted exothermic curve is assigned to the reduction of liberated Fe_2O_3 to lower oxides of iron, e.g. Fe_3O_4/FeO. Here, it is to be noted that the reduction of iron oxide occurs mainly due to the liberation of reducing gases, e.g. CO and H_2 gases during decomposition and oxidation of the organic matter attached with the precipitates.

(iv) The fourth deconvoluted exothermic curve represents reoxidation of Fe_3O_4/FeO to γ-Fe_2O_3. This distinct exothermic peak lies in the range between 257 and 316 °C. During this oxidation, there always remains a probability of α-Fe_2O_3 formation. Possibilities of the incomplete oxidation of Fe_3O_4/FeO to γ-Fe_2O_3 is also not ruled out. Hence, γ-Fe_2O_3 may well contain small amount of α-Fe_2O_3 and Fe_3O_4. This implies that after heat treatment of the powder precipitates in this temperature range may yield nanoparticles comprising γ-Fe_2O_3, α-Fe_2O_3 and Fe_3O_4.

(v) The fifth deconvoluted broad curve can be attributed to the agglomeration and growth of evoluted nanosized iron oxide particles. This also implies that the heat treatment temperature and time will exert a significant influence on the physical as well as chemical characteristics of the nanoparticles.

Fractional conversion (α) values of the individual deconvoluted curves at different heating rates were evaluated and are shown against temperature in Figs. 5.4, 5.5, 5.6, 5.7, 5.8.

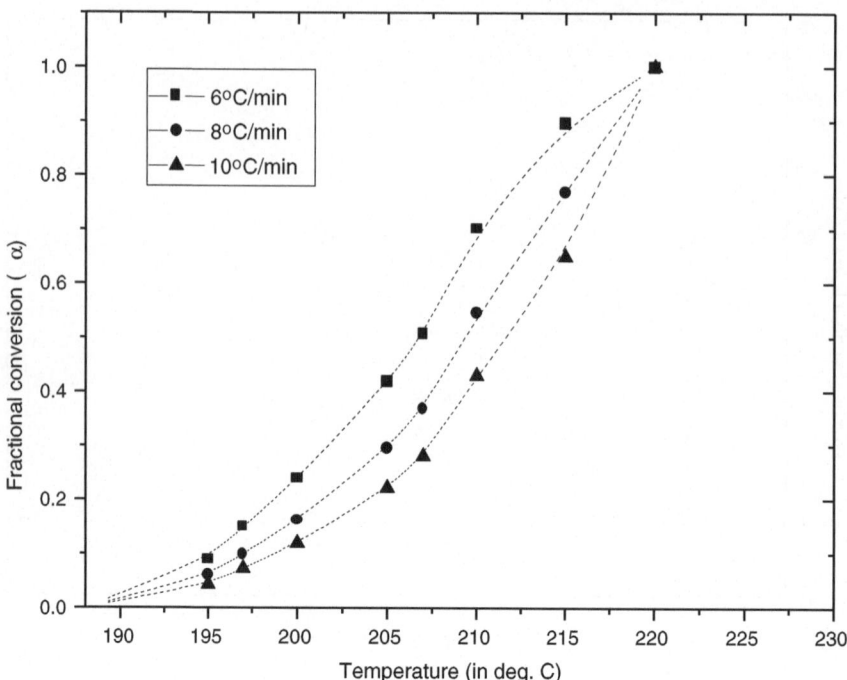

Fig. 5.4 Effect of heating rates on the non-isothermal kinetics of first exothermic deconvoluted pattern

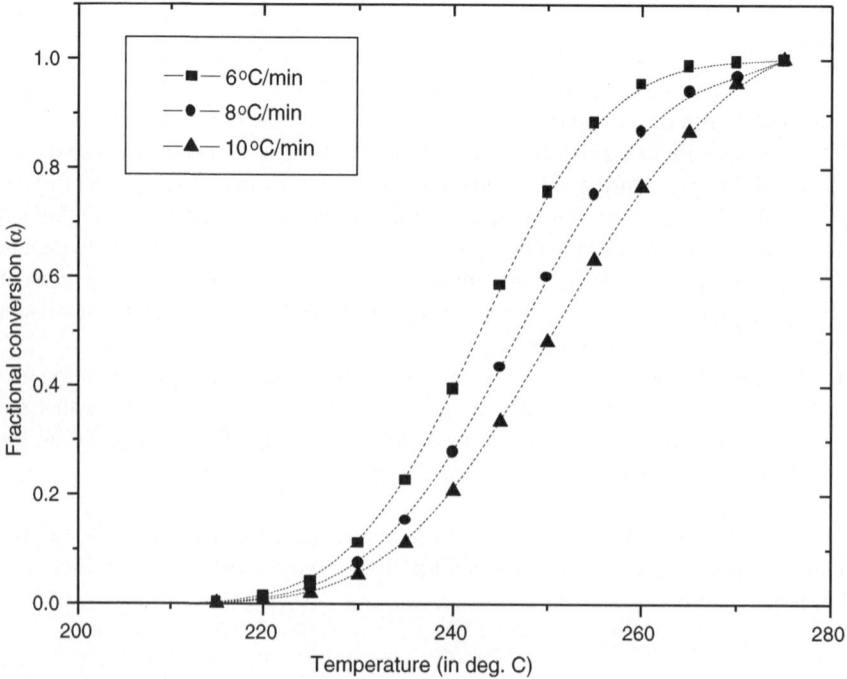

Fig. 5.5 Effect of heating rates on the non-isothermal kinetics of second exothermic deconvoluted pattern

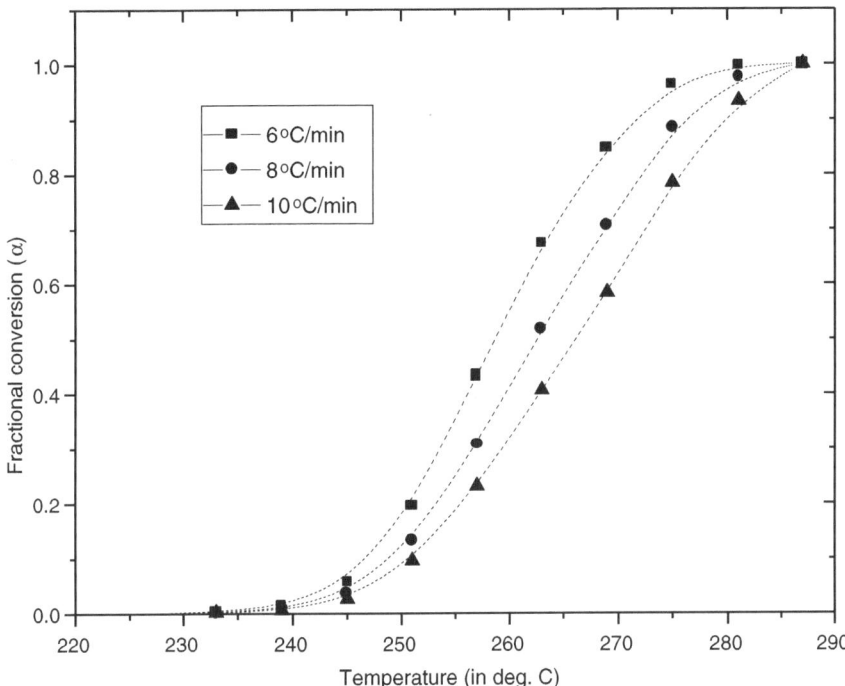

Fig. 5.6 Effect of heating rates on the non-isothermal kinetics of third exothermic deconvoluted pattern

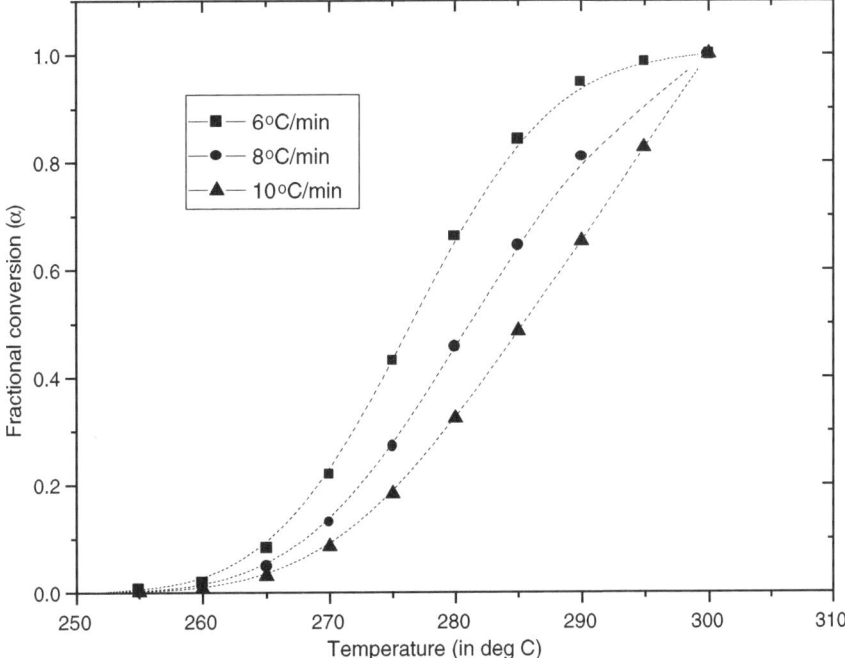

Fig. 5.7 Effect of heating rates on the non-isothermal kinetics of fourth exothermic deconvoluted pattern

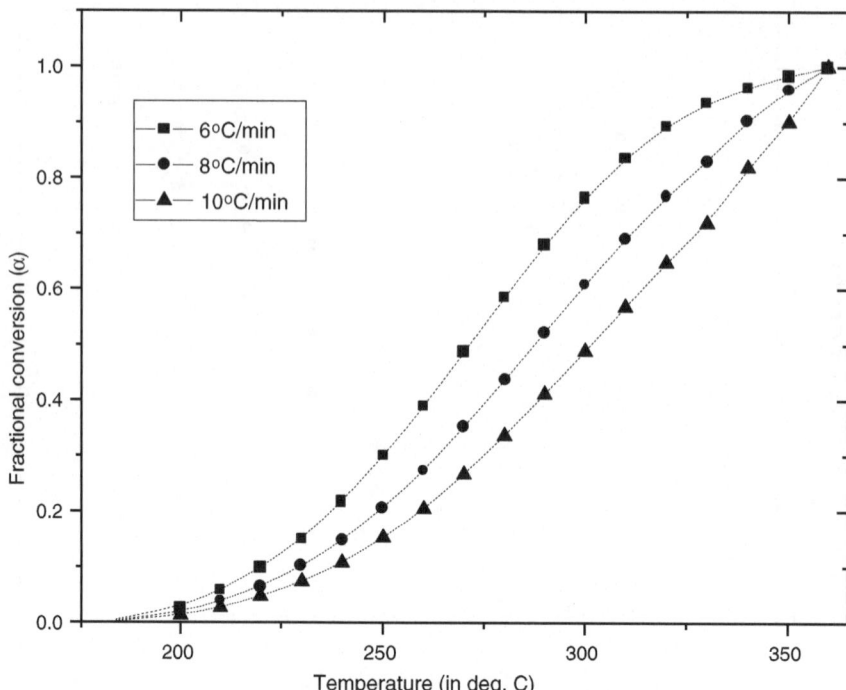

Fig. 5.8 Effect of heating rates on the non-isothermal kinetics of fifth exothermic deconvoluted pattern

Figures 5.9, 5.10, 5.11, 5.12, 5.13 exhibit the activation energy plots of the deconvoluted patterns at different heating rates.

Analysis of the non-isothermal kinetic data of the individual deconvoluted DSC patterns reveals that the activation energy of the reactions represented by the first deconvoluted pattern requires highest activation energy for commencement of the reaction. Therefore, the reactions, which are represented by the above curve can be considered as the slowest and therefore rate controlling. It is interesting to note that the kinetic law, which remains operative for all the five types of reactions, is nucleation and growth type (NG1, NG2). Therefore, it seems to be quite obvious that the overall reaction mechanism for the evolution of iron oxide nanoparticles should follow nucleation and growth type mechanism. It is also interesting to note that the fifth deconvoluted DSC pattern representing growth of the evolved nanoparticles requires very low activation energy (34 kJ/mol). Such low activation energy is typical of diffusional processes. Therefore, the above findings suggest that the growth of the evolved nanoparticles occurs by the diffusion of ions.

On the basis of the above findings, a set of temperatures were identified for performing heat treatments on the sample to corroborate the identified mechanisms for evolution of nanoparticles. The sample was heat treated at 200, 250, 300 and 350 °C without any holding time. The desired temperature of the furnace was maintained by a PID controller with an accuracy of ±1 °C. X-ray diffraction

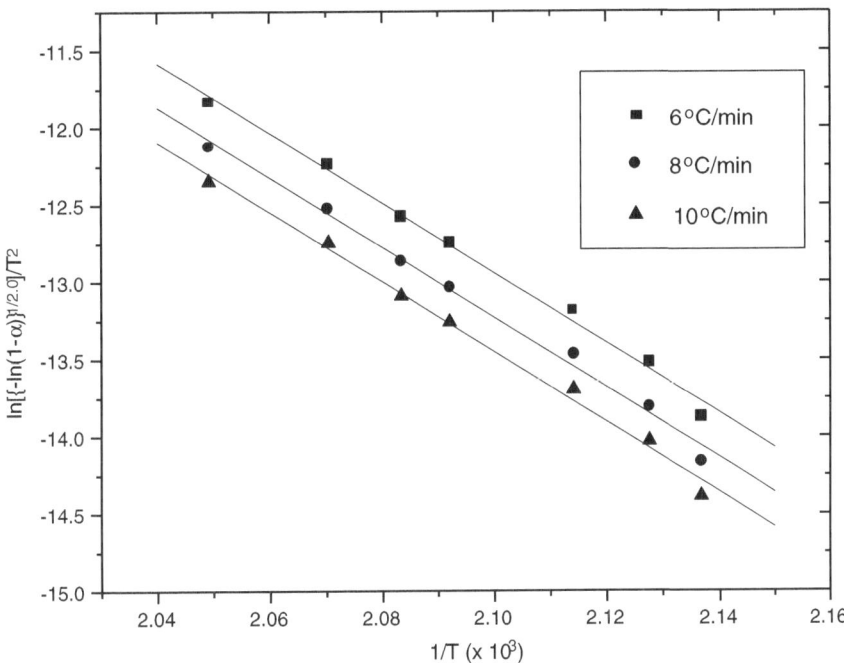

Fig. 5.9 $Ln[g(\alpha)/T^2]$ versus. $1/T$ plots for the first deconvoluted pattern for different heating rates. The activation energy of the reaction, evaluated from the slope of the lines, is 188 kJ/mol

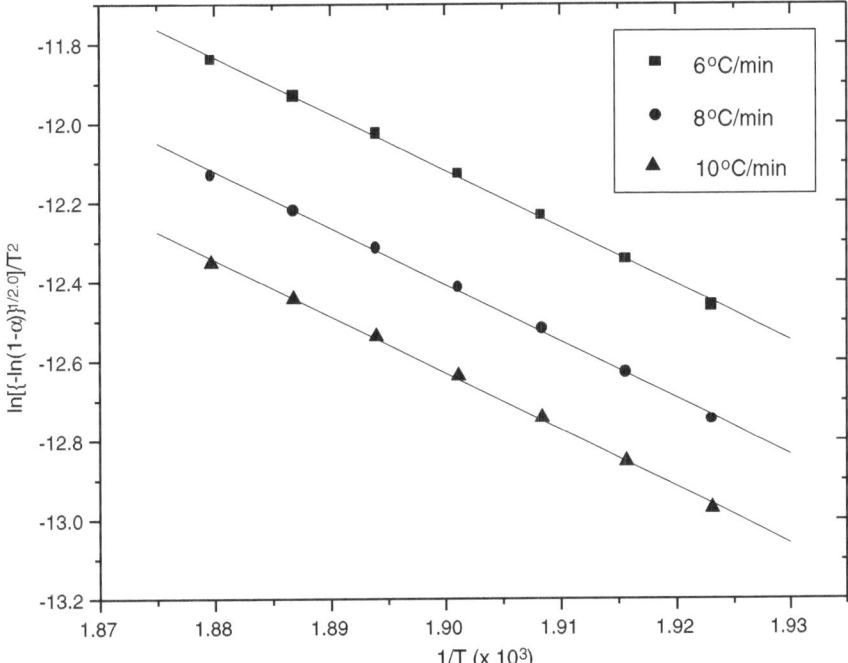

Fig. 5.10 $Ln[g(\alpha)/T^2]$ versus $1/T$ plots for the second deconvoluted pattern for different heating rates. The activation energy of the reaction, evaluated from the slope of the lines, is 118 kJ/mol

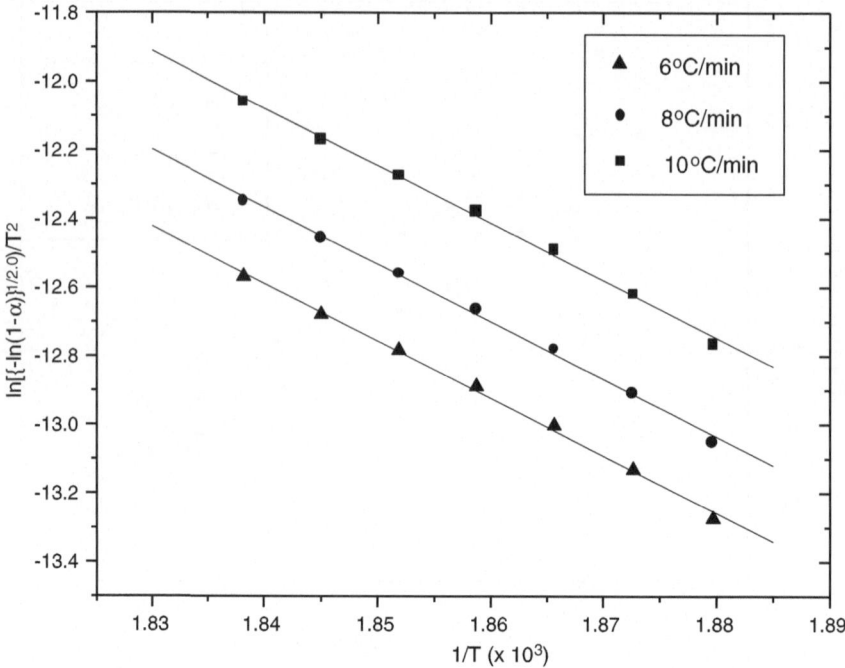

Fig. 5.11 $Ln[g(\alpha)/T^2]$ versus $1/T$ plots for the third deconvoluted pattern for different heating rates. The activation energy of the reaction, evaluated from the slope of the lines, is 139 kJ/mol

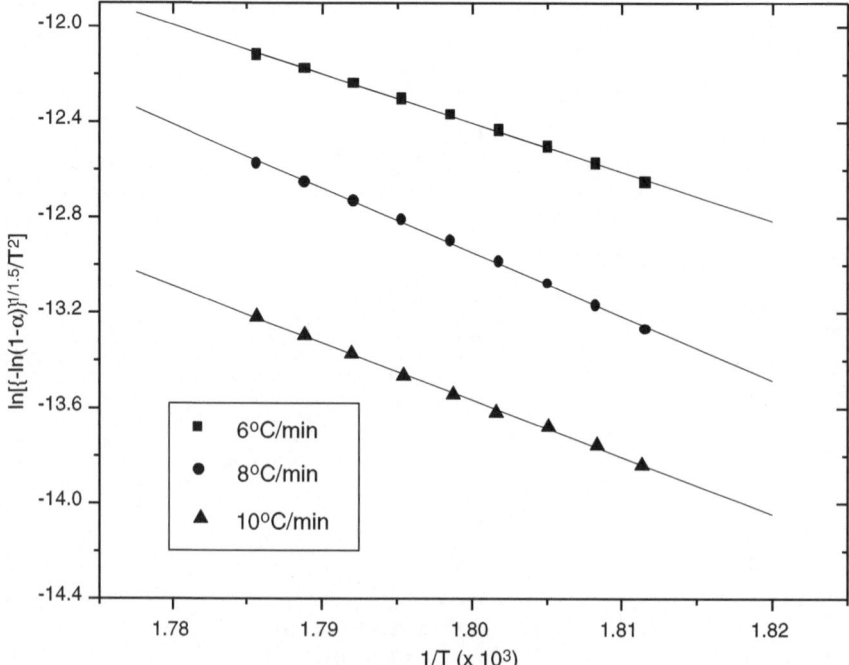

Fig. 5.12 $Ln[g(\alpha)/T^2]$ versus $1/T$ plots for the fourth deconvoluted pattern for different heating rates. The activation energy of the reaction, evaluated from the slope of the lines, is 170 kJ/mol

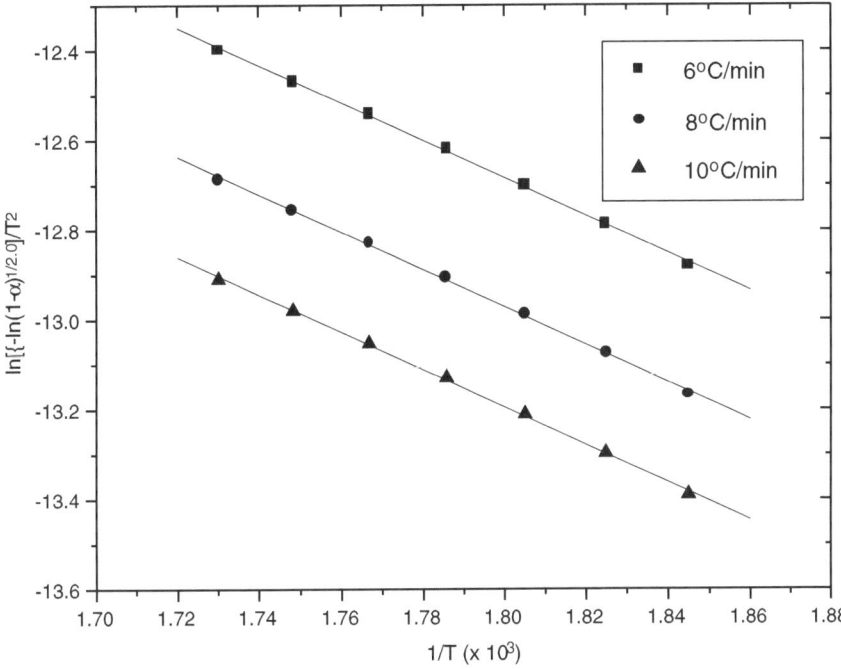

Fig. 5.13 Ln[$g(\alpha)/T^2$] versus $1/T$ plots for the fifth deconvoluted pattern for different heating rates. The activation energy of the reaction, evaluated from the slope of the lines, is 34 kJ/mol

studies were carried out on the heat-treated samples. Figs. 5.7, 5.8, 5.9 show the XRD patterns of these heat-treated samples. The observed asymmetry in the most prominent peak of the XRD patterns of the heat-treated samples can be attributed to the presence of superimposed characteristic peaks of multiple phases. With the objective to identify the phases, the 2θ region concerned with the most prominent peak was subjected to a carefully conducted deconvolution exercise. The deconvolution was preceded by plotting $d^2I/d\,(2\theta)^2$ as a function of 2θ (I being the relative intensity of the peaks) within the said 2θ region to identify the locations of the superimposed peaks. The results of the deconvolution analysis for the most prominent peak for each pattern are shown in the inset. Here, it would be worthwhile to mention that the characteristic peaks of the expected iron oxide and FeOOH phases possessing maximum intensity lie within the 2θ range of the most prominent peak of the XRD patterns.

The XRD profile analysis of heat-treated samples exhibits the presence of various transient phases, identified based on standard ASTM d_{hkl} values for all the peaks. The deconvoluted XRD pattern (Fig. 5.14) of the sample heat treated at 200 °C indicates the presence of multiple metastable phases of iron oxide and Fe-O-OH system. Analysis of the XRD patterns of the sample heat treated at 250 °C reveals that they contain a mixture of Fe_3O_4, γ-Fe_2O_3 and α-Fe_2O_3. The XRD pattern analysis of the sample heat treated at 300 and 350 °C suggests that it

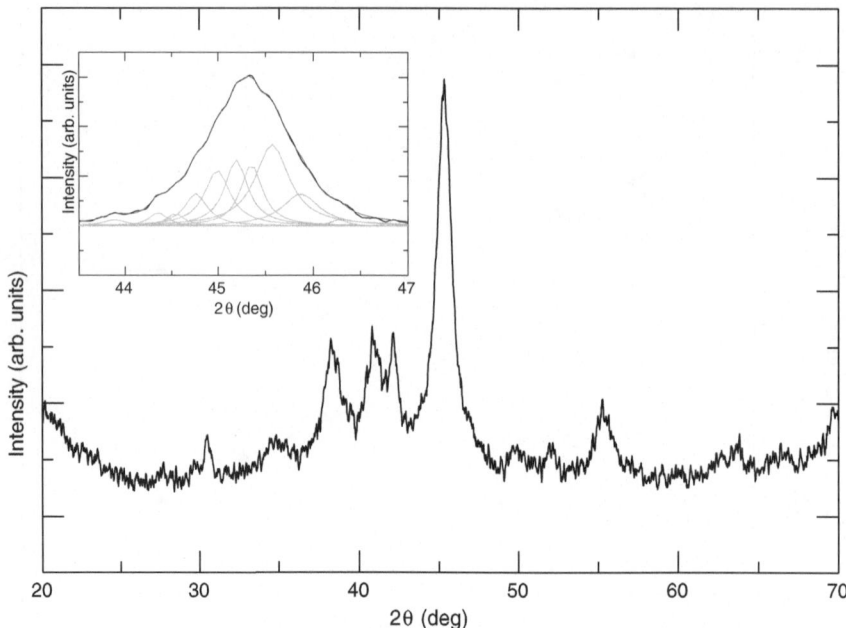

Fig. 5.14 XRD pattern of sample heat treated at 200 °C. The inset shows the deconvoluted main peak

contains predominantly γ-Fe_2O_3. The above phenomena may be explained on the basis of the fact that during sample preparation prior to heat treatment, the attached organic matter generates CO and H_2 by decomposition, which are both reducing gases. The above gases will reduce Fe_2O_3 and Fe-O-OH system to lower oxides of iron, mainly Fe_3O_4. However, since the processing temperature is low, the reduction potential of these gases is also low. Therefore, under the prevailing processing conditions, the Fe_2O_3 and Fe-O-OH system is expected to get reduced partially to lower oxides of iron. However, the evolution of reducing gases ceases with the completion of decomposition reaction. Hence, the iron oxides reoxidize once again under the prevailing conditions to γ-Fe_2O_3. Therefore, it would be quite logical to expect that the samples will contain γ-Fe_2O_3 along with the crystallographic variants of iron oxides. The samples with varying heat treatment temperature are also leading to the growth of the particles and cause evolution and disappearance of crystallographic phases of iron oxide system. This XRD analysis results are corroborating with the conclusions of kinetic analysis of the intermediate reactions (Figs. 5.15, 5.16, 5.17).

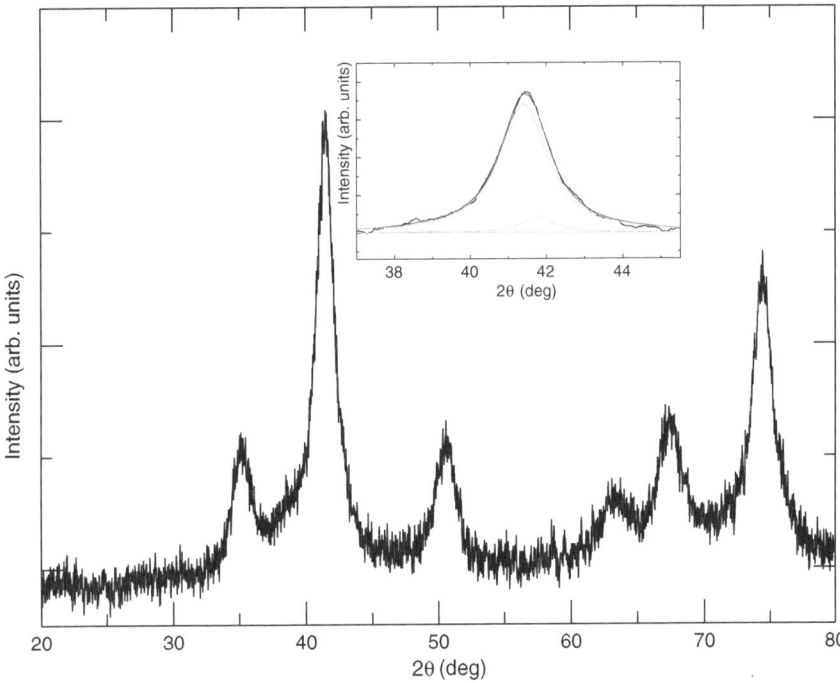

Fig. 5.15 XRD pattern of sample heat treated at 250 °C. The inset shows the deconvoluted main peak

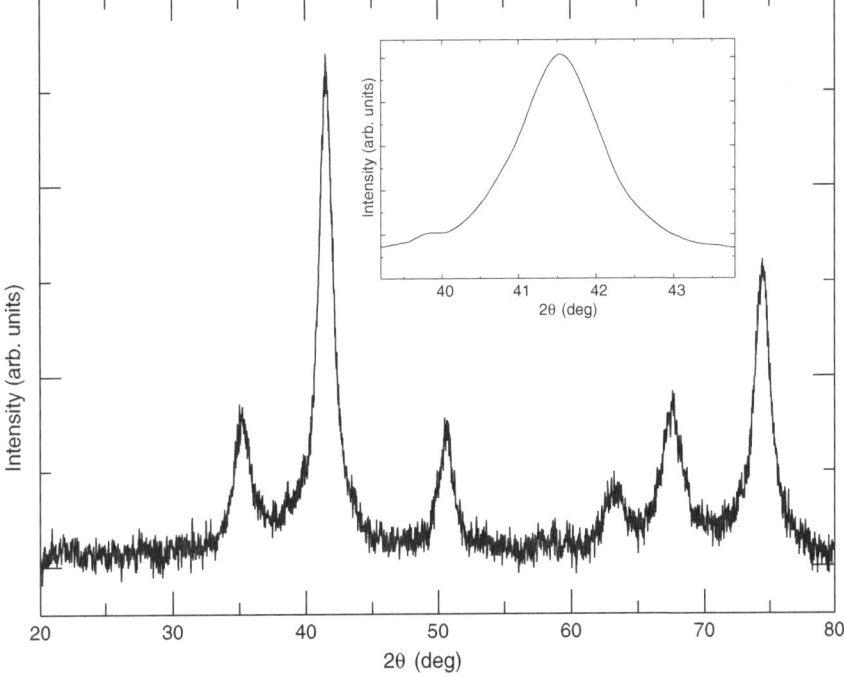

Fig. 5.16 XRD pattern of sample heat treated at 300 °C. The inset shows the deconvoluted main peak

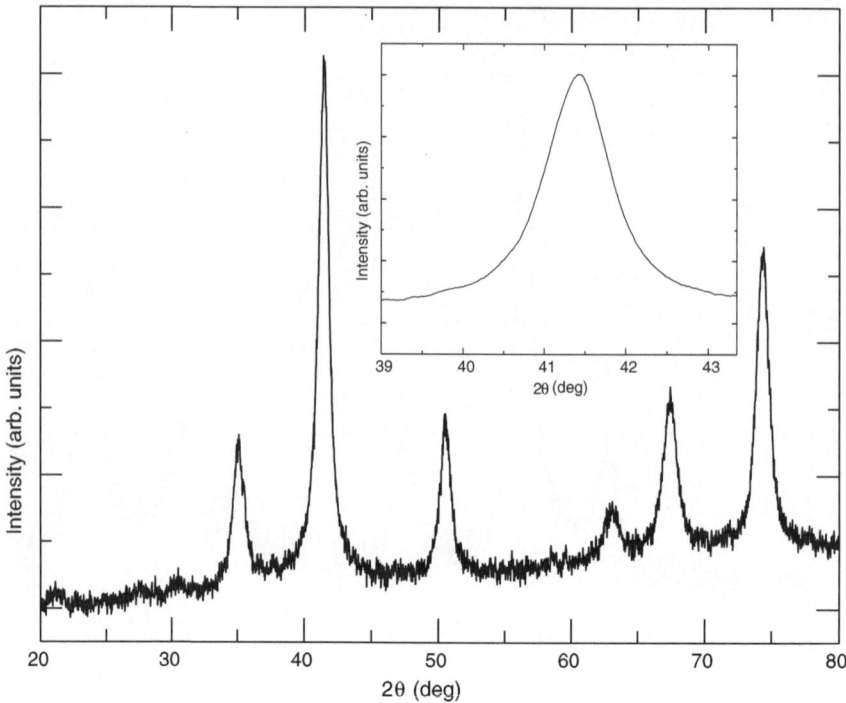

Fig. 5.17 XRD pattern of sample heat treated at 350 °C. The inset shows the deconvoluted main peak

References

1. Buriak JM (2004) Nat Mater 3:847
2. Dorvee JR et al (2004) Nat Mater 3:896

Chapter 6
Summary

The present non-isothermal kinetic analysis method successfully identifies the mechanism of the solid state process. The activation energy can be calculated more accurately from the kinetic data. The apriori knowledge of the reaction mechanism is not at all required. Hence the kinetic laws are ascertained directly from the non-isothermal kinetic data. Even kinetic law(s) at any temperature that fall within the experimental range can be ascertained. This newly developed method has been found to be effective in identifying the kinetic law and thereby the mechanism of iron oxide nanoparticles evolution from an organic precursor. The intermediate reaction in the evolution process of iron oxide nanoparticles has been identified by deconvoluting the kinetic data. The rate controlling step has also been identified on the basis of the evaluated activation energy. The kinetic law representing the overall reaction as well as the intermediate reactions has been found to be matching. The subject matter demonstrated in this book has clarified that solid state kinetic analysis has a theoretical physical meaning and is not merely based on goodness of data fits to complex mathematical expressions.

P. Deb, *Kinetics of Heterogeneous Solid State Processes*, SpringerBriefs in Materials, DOI: 10.1007/978-81-322-1756-5_6, © The Author(s) 2014

About the Author

Pritam Deb is currently Associate Professor at Tezpur University (Central University), India and also holds concurrent position of Max Planck Fellow at the Max Planck Institute for Eisenforschung, Germany. He received his Ph.D. from Jadavpur University. Dr. Deb is a recipient of several national and international accolades, which include DAE Young Scientist Research award, ISCA Young Scientist award, MRSI Young Scientist award and BHUMET Golden Jubilee fellowship. Dr. Deb has published over 40 peer-refereed research papers and delivered over 25 invited talks at various conferences during the last 5 years. In addition, three patents and two software copyrights are also credited to him. He has published one book with McGraw Hill and edited one special volume of Indian Journal of Physics.

P. Deb, *Kinetics of Heterogeneous Solid State Processes*, SpringerBriefs in Materials,   47
DOI: 10.1007/978-81-322-1756-5, © The Author(s) 2014

About the Book

Kinetic studies have traditionally been extremely useful in characterizing several physical and chemical phenomena in organic, inorganic and metallic systems. It provides valuable qualitative, quantitative and kinetic information on phase transformations, solid state precipitation, crystallization, oxidation and decomposition. Unfortunately, no single reference comprehensively presents non-isothermal kinetic analysis method for the study of complex processes, determining the actual mechanism and kinetic parameters. This book provides a new method for non-isothermal kinetics and its application in heterogeneous solid state processes. In the backdrop of limitations in existing methods, this book presents a brief review of the widely used isothermal and non-isothermal kinetic analysis methods.

P. Deb, *Kinetics of Heterogeneous Solid State Processes*, SpringerBriefs in Materials, 49
DOI: 10.1007/978-81-322-1756-5, © The Author(s) 2014